Essentials of Electricity

By William H. Timbie

Elements of Electricity, Fourth Edition
Basic Electricity for Communications, Second Edition
    BY WILLIAM H. TIMBIE AND FRANCIS J. RICKER
Essentials of Electricity, Third Edition
    BY WILLIAM H. TIMBIE AND ARTHUR L. PIKE
Industrial Electricity
  Volume I: Direct-Current Practice, Second Edition
  Volume II: Alternating-Current Practice
    BY WILLIAM H. TIMBIE AND FRANK G. WILLSON

By William H. Timbie and Henry H. Higbie

Essentials of Alternating Currents, Second Edition

By William H. Timbie, Vannevar Bush, and George B. Hoadley

Principles of Electrical Engineering, Fourth Edition

Essentials
of Electricity

THIRD EDITION

by the late William H. Timbie

Revised by Arthur L. Pike
ASSOCIATE PROFESSOR, TUFTS UNIVERSITY

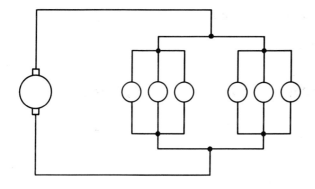

John Wiley and Sons, Inc., New York and London

Copyright 1913, 1931, and 1958
by William H. Timbie
Copyright © 1963 by John Wiley & Sons, Inc.
All rights reserved.
This book or any part thereof
must not be reproduced in any form
without the written permission of the publisher.

Library of Congress Catalog Card Number: 63-8053
Printed in the United States of America

Preface

Since the late Professor Timbie was able to review only the first portion of my revision of his book, I have had to decide many questions that affect the nature of this third edition by myself. Because he left an impressive record of unusually effective communication in his writing, I hope that my changes have preserved the spirit of his style.

Yet changes had to be made. If the fundamental theory of electrical science seemed in 1931 to Professor Timbie to have undergone "decided change," he must surely have felt that the decades since then have unfolded a major revelation about conduction in materials in the solid state. Many industrial applications of great financial value between 1912 and 1931 have now been replaced in importance by other kinds of applications, especially those that relate to a-c systems, communication of time-varying signals, and automatic control. Furthermore, the student in this decade has a different background from the student Professor Timbie knew. Today's student uses television and direct-dialing telephones (to name evident examples of complicated electrical devices) with at least the vague understanding of familiarity. Therefore, choices concerning the scope, order, and level of topics had to be made for this introductory textbook.

My decision was that emphasis should rest on d-c systems and devices. In circuits with time-invariant voltages it seemed possible to present rather complicated numerical problems without expecting the reader to have any formal background in mathematics, even algebra. In this way, Professor Timbie's verbalized forms of Ohm's law and of some network-reduction schemes have been retained.

I believe that Professor Timbie understood clearly the need for motivating the student and the necessity that the teacher organize his presentation carefully in a well-planned, programmed sequence. In the light of my own teaching experience, I have changed the order of presenting topics while retaining the plan "to build outward, and by very small steps, from the starting point of the reader's own experience in the commercial shop or the school laboratory," as Professor Timbie wrote in his first preface. As an example of this change, I have discussed meters early so that they can be an aid in learning about the behavior of complicated circuits. In order that the reader will understand the forces behind the indications of a meter, I have placed a discussion of magnetism and magnetic forces just before the chapter on meters.

Because I believe that circuit analysis forms the most secure basis for an introduction to electrical applications, I have substituted material on circuit analysis for certain sections that were deleted. In particular, I have tried to use applications as a medium to introduce new topics in circuit analysis, where possible. The two Kirchhoff's laws and the calculations for power have been presented as a general foundation for further study, although still in verbalized form. Extensive additions were made to the chapter on batteries so that the general concept of internal resistance of a source could be developed from the special need for its consideration with battery cells. My purpose in introducing a verbalization of the Thevenin method of network reduction was twofold: to present an important topic in circuit analysis, and to provide a framework for subsequent presentation of the equivalent circuit for the armature of a d-c machine.

Although many of the old problems appear in this new edition, practically all have been renumbered and revised. Changes have been made in the numerical values quoted to give emphasis to changes in commercial practice. In place of the many problems deleted are some new ones covering the same concepts, and there are also many entirely new problems in the areas covered in the new sections of the text. There has been a small increase in the total number of problems over that of the previous edition.

With the elimination of electronic devices and their time-varying signals from this book, the need to discuss electron motions was questioned. Although the temptation is strong to define current as the motion of electrons for electron tubes, this departure from the older view of electrical theory introduces two new difficulties. First, the motions of positively charged holes in semiconductors cause exactly the same problem with the electron-current concept as the motions of electrons cause with the conventional-current concept. Second, unless the direction of magnetic flux lines is defined as positive when leaving a south pole, all of the geometry of electromagnetic field theory must be changed to a left-handed set of coordinates when current is defined as electron current. When the student goes on to advanced work, he will find that almost all of electrical theory is presented in terms of right-handed coordinates and in terms of positive current as the positive velocity of positive charge. Of course this viewpoint gives no difficulty: negative charges (or electrons) with negative velocities constitute a positive current. Furthermore, this viewpoint provides a unified basis for study of filter circuits, electric machines, antennas, waveguides, traveling-wave tubes, and semiconductor devices, as well as what is now classical electron-tube theory. Therefore, I have emphasized conventional current in this book with the hope that later algebraic treatments can build on this material directly without a new formulation. I recognize my partisan statement of this matter.

In the preparation of this edition I feel that I have drawn upon, in uncounted ways, my associations and experiences both at the Massachusetts Institute of Technology and at Tufts University. In the Departments of Electrical Engineering in both universities, assignments of technical writing and editorial work have provided me with diverse points of view about the problems of teaching through written material. And it shall not go without saying that my greatest encouragement and assistance have come from my wife.

ARTHUR L. PIKE

Arlington, Mass.
November 1962

Contents

Chapter 1 Introduction 1

Electricity at Work 1
Charge; Coulomb 1
Current; Ampere 1
Work; Potential 2
Joule; Volt 3

Chapter 2 Ohm's Law 6

Proportionality; Ohm's Law 6
Resistance; Ohm 7
Ohm's Law; Voltage 9
Ohm's Law; Current 9
Symbols 11
Electrons; Conventional Current 12
Polarity; Current Direction 13
Ideal Battery 14

Chapter 3 Simple Electric Circuits 17

Definition of Series and Parallel Circuits 17
Series Circuit; Current 18
Series Circuit; Potential 18
Series Circuit; Resistance 20
Series Circuit; Current, Voltage, and Resistance 22
Application of Ohm's Law 23
Parallel Circuit; Voltage 26
Parallel Circuit; Current 27
Parallel Circuit; Resistance 28

x   Contents

Chapter 4   Magnets and Magnetism — 38

Magnetic Principles   38
Flux Lines   40
The Compass Needle   42
Electromagnets—The Right-Hand Rule   44
Forces of Magnetism   47
Current-Carrying Wire   49

Chapter 5   Meters and Instruments — 55

Definitions   55
Convenient Units   56
The Milliammeter   56
The Ammeter   60
The Voltmeter   63
Recording Meters   69
Measuring Resistance   71
Iron-Vane Instruments   76

Chapter 6   Series-Parallel Circuits — 79

Circuit Connections   79
Kirchhoff's Current Law   80
Kirchhoff's Voltage Law   80
Voltage Diagrams   83
Parallel Loads on a Line   85
Ladder Groupings   89
Line Drop for Fixed-Current Loads   92
Source Voltage for Fixed-Current Loads   94
Ladder-Circuit Voltages for Fixed Load Resistance   96
Ladder-Circuit Voltages with Fixed Source Voltage   99
The Unit-Ampere Method   103

Chapter 7   Electric Power and Energy — 111

The Time Rate of Doing Work   111
Units of Power   112
Three Forms of the Power Equation   114
Measurement of Power in an Electric Circuit   115
The Wattmeter   118
Line Loss   119
Kilowatt and Horsepower   121

Efficiency of Electric Devices   122
Work and Energy; Kilowatthour   124
The Watthour Meter   124

Chapter 8   Batteries   131

Generators versus Batteries   131
Battery Terminology   132
Internal Voltage   133
Internal Resistance   133
Terminal Voltage   135
Cells in Series or Parallel   138
Unlike Cells   142
Equivalent Circuits   145
The Thevenin Method   147
Series-Parallel Cell Groupings   151
Dry Batteries   154
Electrolysis and Electroplating   156
Electrotyping   158
Refining of Metals   158
Electrolytic Destruction   158
Automotive Batteries   159
Rating of Batteries   162
Care of Automotive Batteries   165
Specialized Industrial Batteries   166

Chapter 9   Wire and Wiring Systems   176

Insulators and Conductors   176
Resistance Increases with Length   177
Resistance Decreases as Area Increases   177
Circular Wire   180
Circular Mil; Circular Mil-Foot   182
Copper Wire   183
Line Drop and Dimensions   185
Wire Table for Copper   187
Stranded Wire   191
Other Conductors   192
Safe Current-Carrying Capacity for Copper Wires   193
Voltage Level and Line Loss   194
Three-Wire System   197
Unbalanced Three-Wire System   198
Voltage Diagrams for Unbalance   199
Open Neutral   201

xii   Contents

Chapter 10   Switches and Signal Devices   206

Switches, Relays, and Contactors   206
Basic Contact Connections   208
Electric Bells   211
Buzzers   213
Electric Lock Opener   214
Annunciators   214
Control of Incandescent Lamps   217

Chapter 11   D-C Machines   223

Reversibility   223
Basic Principles; Motor Action and Faraday's Law   223
Basic Parts of Electric Machines   225
Magnetic Paths   229
Types of Field Coils   230
Types of Connections   232
Number of Brushes   233
Commutating Poles   233
Equivalent Circuits for Basic Parts   236
Separately Excited Generator   237
Shunt-Generator Connections   237
Shunt-Generator Build-Up   239
Compound-Generator Connections   241
Torque and Power   243
Energy Conversion   245
Shunt Motor   248
Starting Resistance   251
Manual Starting Boxes   254
Automatic Starters   260
Series Motors   264
Some Precautions for Operating Shunt and Series Motors   265
Some Basic Machine Troubles   266
Sparking at Brushes   268
Noise   271
Hot Bearings   272
Hot Field Coils   273
Hot Commutator   274
Hot Armature Coils   274
Generator Fails to Build Up   275
Voltage of Generator Too Low   277
Generator Voltage Too High   277

Motor Fails to Start   277
Motor Speed Too High   278
Motor Speed Too Low   279
Speed Control of Shunt Motors   279
Speed Control with Separate Excitation   280
Control Generators   280
Conclusion   281

Appendix   291

Index   295

1
Introduction

1. Electricity at Work. Electricity, in motion, lights lamps, drives motors, actuates communication and signaling devices, and decomposes chemical compounds. It produces heat in electric furnaces and in many types of smaller appliances such as soldering irons, stoves, heating pads, and cigarette lighters in automobiles. Electricity, at rest, has fewer and less familiar practical applications. Accordingly, our study here is confined largely to electricity at work, that is, in motion.

2. Charge. Coulomb. When we say "electricity," we mean a certain kind of substance, called *charge*. This substance, in various arrangements, has been found to compose all of the materials we know. Wires, like all metals, have some of their charge arranged so that it can be made to move upon application of the proper kind of pressure. Electricity in motion, then, is charge moving in wires, somewhat as water flows in pipes. Quantities of charge are measured in terms of a unit called a *coulomb*, similar to the way quantities of water are measured in gallons. Although the coulomb is not the smallest unit of charge (see Section 2-6), it is a convenient electrical unit because it is related to other electrical units by simple numbers. The important idea to grasp is that the flow of electricity means charge in motion in a wire.

3. Current. Ampere. When water is flowing through a pipe we never ask, "How much water is in the pipe?" but rather "At what rate is water

2 Essentials of Electricity

flowing through the pipe?" That is, "How much water flows through the pipe in a given time?" The answer would not be "Five gallons," but "Five gallons per second." We are interested not in the quantity, but in the quantity that passes through in a second.

Similarly, we never ask, "How much electricity is in that wire?" but rather, "What current exists in that wire?" By this question we mean, "How much electricity flows along that wire per second?" The answer could not be "Five coulombs," but it might be "Five coulombs per second." We are interested not in the quantity, but in the quantity that passes in a second.

It is unfortunate that we have no name for the unit flow of water, which means 1 gallon per second. Consequently we always have to say 5 gallons per second, 10 gallons per second, etc.

But we are fortunate in not having to use the term "coulomb per second" to denote quantity of electricity that flows per second. We call a *coulomb per second* an *ampere*. Instead, therefore, of answering the above question by "Five coulombs per second," we would say "Five amperes." Thus we do not need to say "per second" each time, as "amperes" means "coulombs per second." So 10 amperes means 10 coulombs per second, etc.

Since we are always concerned with the rate of flow of electricity and not the quantity, we continually employ the term *ampere* and rarely use the term *coulomb*.

Thus an ordinary incandescent lamp, with a tungsten filament, when glowing normally, may take a current of $\frac{1}{2}$ ampere; that is, $\frac{1}{2}$ ampere or $\frac{1}{2}$ coulomb per second is the rate of charge flow through the filament.

4. Work. Potential. Although charge flowing in a wire acts similar to water when flowing in a pipe, the comparison depends on thinking about a pipe that is always filled completely full of water. A piece of wire not connected to any kind of electric circuit is, by its very nature, already filled full of charge. When the wire is part of an electric circuit and a current is made to exist in the wire, just as much charge leaves one end as enters at the other end. Therefore, the wire remains completely filled with charge. However, the charge flowing in the wire must thread its way through a complicated atomic obstacle course. Its maneuvers require that, somewhere in the circuit, there must be a source of work that can push the charge over the atomic hurdles.

What do we mean by *work?* Work is a technical term for energy expended. In mechanics, work is often measured in foot-pounds. For example, if a man has to push continually with a force of 20 pounds to slide a crate 10 feet across a floor, we say that he has done 200 foot-pounds of work. Notice that force and distance are both required. If he pushes with a force of 20 pounds on another crate and it does not move, he has not done any *work* in the way we use the word technically. However, since he could move the first crate, we say that it is possible for him to do 200 foot-pounds of work, or that he has a potential work of 200 foot-pounds. Thus, the possibility of doing work, whether or not any is done, is called potential work, potential energy, or often simply *potential*.

Recall that when current exists in a wire there must be a source of work in the electric circuit. This source is usually a battery or an electric generator. When the source causes current to exist in a circuit joined to the two terminals of the source, work is done. However, since the source has the possibility of such work even when nothing is connected to the terminals, it is necessary to have a way of describing the *potential* of the source. For electrical work, metric units rather than British units are used.

5. Joule. Volt. In the metric system, the unit of work is the *joule*. Like the British unit of foot-pound, the joule is the product of a force and a distance; however, it is the special name for the metric product, in the same way that an ampere is the special name for 1 coulomb per second. One joule equals 0.738 foot-pound, or almost $\frac{3}{4}$ of a foot-pound.

For comparison with the man pushing the crate, suppose that an electric source can do 200 joules of work when it forces charge through a circuit. Suppose further that this source moves 10 coulombs of charge while doing 200 joules of work. Now if we were going to compare one source with another, we would rather know the joules of work that could be exerted on just one coulomb. Thus, we would say that this source could exert 20 *joules per coulomb* of charge moved. Again in electrical units there is a shorter, special name: one joule per coulomb is called a *volt*. Therefore we would say a source that exerts 20 joules per coulomb has a potential of 20 volts.

Many books and many electrical workers use a name for potential that is formed from the name of the unit of potential. This name for potential is called *voltage*. Applying this usage to the previous paragraph, we would say a source that exerts 20 joules per coulomb has a voltage of 20 volts. Since both the names of *voltage* and *potential* are in common use, it is important to know both. In this book we use both names.

Be careful to notice that the previously mentioned potential of 20 volts (20 joules per coulomb) exists in the source whether or not something is connected. If a complicated circuit of many wires is connected

4 Essentials of Electricity

to the source, the actual amount of work done depends on the number of coulombs of charge moved. However, we will still find that the actual work is related to the potential of 20 volts. Notice also that no distance is involved with electric charge. Therefore, to compare an electric circuit with a system of filled water pipes, we might think of voltage as pressure on the water, but not as force.

If we keep in mind the idea of filled pipes, we can form another idea about electric potential. When a section of filled pipe has a current of water in it, there must be a difference in pressure between the two ends in order to force the water to move. In the same way, when a wire has an electric current in it, there must be a difference in potential between the two ends to force the charge to move. Since electric potential is measured in volts, this difference in potential between the two ends of the wire is also measured in volts. If we find that there is a potential difference (or simply potential) of 20 volts between the two ends of a wire, we mean that 1 coulomb of electric charge will have to have work of 20 joules performed on it, if it is to be moved from one end of the wire to the other.

It is important to understand amperes as current and volts as potential at the beginning. Such clarification will help to avoid the following misconception: When the statement is made that across the terminals of a switch are 110 volts, we often hear an uninformed person ask, "How many amperes are there?" If the switch is in the OFF position, there is no current and of course there are no amperes, no more than there would be current in a pipe if the valve were turned off. Even though there was great potential energy in the pressure of the water in a closed pipe we would not ask how many gallons were passing in a second, knowing that such a current depends on whether or not we turn on the valve. Even then the gallons per second depend upon what is connected into the pipe line. So in an electric circuit, even if the potential in volts is known, the current depends upon whether or not the switch is turned ON, and also upon what is connected into the circuit for the charge to flow through.

Again, we read newspaper accounts of an accident in which a man was injured by so many volts passing through his body. This statement is not correct. The volts merely caused a certain current (amperes) to exist in the person's body and this current injured the man. These examples are given so that the student can, at the beginning, get a clear understanding of the meaning of *volt* and *ampere* and can avoid using them incorrectly. Usually, no mistakes will be made if we say "volts across" and "amperes in" a circuit.

SUMMARY

ELECTRICITY may be considered to flow along a wire conductor, much as water flows through a pipe.

CHARGE is the substance of ELECTRICITY, and it makes up all materials. The QUANTITY of CHARGE is measured in COULOMBS.

CURRENT is the rate of flow of CHARGE. CURRENT is measured in AMPERES, expressing the QUANTITY of CHARGE that passes a point IN ONE SECOND.

WORK is the product of force and distance. In British units, WORK is measured in foot-pounds; in metric units, WORK is measured in JOULES.

POTENTIAL is the WORK, either possible or actual, that is exerted on a unit of CHARGE if that CHARGE is caused to move from one end of a wire to the other. POTENTIAL is measured in VOLTS, expressing the WORK per unit CHARGE in JOULES PER COULOMB.

Electrical Quantity	Units of Measure
CHARGE	COULOMBS
CURRENT	AMPERES
POTENTIAL	VOLTS

2
Ohm's Law

1. Proportionality. Ohm's Law. In Chapter 1 we found that current is measured in amperes and that potential is measured in volts. Recall that potential is the work per coulomb across a given section of an electric circuit when work is done in that section. If we should compare two different sections, one made of copper wire and the other made of iron wire, we would expect the work to be different even when the same number of coulombs passes through both sections in the same time. Although each such section that we might imagine will usually have a different numerical ratio between potential and current (volts and amperes), there is just one important physical law that governs the relation between voltage and current for simple electric circuits. For a truly amazing variety of electric equipment, it has been found by experiment that voltage is *proportional* to current. This important relationship is called *Ohm's law,* after the man who first stated it.

In most electric circuits, voltage is proportional to current.

Although this law seems to be and actually is very simple, we must study it carefully to learn how it applies throughout the entire range of electrical devices.

First, let us see what proportionality means. Suppose there is a long length of copper wire forming a field winding in an electric motor. We will make some measurements of the potential and current for different values of current in this wire. The results of these measurements are:

(1) With a potential of 50 volts, the current is 0.25 ampere.
(2) With a potential of 100 volts, the current is 0.50 ampere.
(3) With a potential of 300 volts, the current is 1.50 amperes.

From the first to the second test, the current was doubled; in order that twice as much current could exist, it was necessary that the voltage be doubled. From the first to the third test, the current became six times larger; the work necessary to maintain the larger charge flow must also be six times larger. The results of the second and third tests should be compared to see that the proportionality still holds.

When two quantities such as the voltage and current of the motor winding are proportional to each other, it is often very helpful to talk about their ratio as a *constant of proportionality*. By this phrase we mean that the ratio of the two proportional quantities is always a constant. Look at the measurements above. If we divide the potential by the current in the first test, we have 50/0.25, or 200. In the third test the ratio of potential to current is 300/1.50, or still 200. Of course we can check that the ratio of the second test also gives 200.

2. Resistance. Ohm. In computing the constant of proportionality for the motor field, we divided the voltage by the current. We could just as well have divided the current by the voltage; the ratio would still be a constant, although 0.005 rather than 200. Whereas in Chapter 3 we look again at the ratio of current to voltage, here we consider only the ratio of voltage to current.

The number we get as a result of dividing voltage by current is important. This ratio or constant of proportionality has a shorter, special electrical name. We say that the ratio of voltage to current in a part of an electric circuit is the *resistance* of that part of the circuit. This ratio represents the opposition or resistance that a source would encounter in attempting to create a current in that part of the circuit. When the ratio is low, we say that the resistance is low. A source with a given potential easily sets up a current in the part having low resistance. The same source could set up only a small current in a part having high resistance, i.e., a high ratio of potential to current.

In honor of the man who first stated the proportional law, the unit of measure for resistance is called the *ohm*. When a potential of 1 *volt* is able to set up a current of 1 *ampere* in a section of a circuit, we say that that section of the circuit has a resistance of 1 *ohm*. Thus, we see that the motor winding above had a resistance of 200 ohms. If we observed

8   Essentials of Electricity

a current of 1 ampere in that winding, we would find that the potential across the winding would be 200 volts.

Of course, actual electrical devices do not always have resistance ratios that are round numbers. For example, a certain incandescent lamp may, when operated with a potential of 120 volts, have a current of 0.833 ampere in its filament. Therefore, the resistance of this lamp will be 120 volts divided by 0.833 ampere, i.e., 120/0.833 or 144 ohms.

Although Ohm's law is complete as we have stated it, we will find that we can solve problems about resistance easily if we write the proportional law another way:

The resistance in which a given voltage will set up a given current equals the quotient of the voltage divided by the current.

These words may be summarized briefly as:

$$\text{resistance} = \frac{\text{voltage}}{\text{current}}$$

or as

$$\text{ohms} = \frac{\text{volts}}{\text{amperes}}$$

All we have done is to state Ohm's law another way.

Example 1. What resistance must an electric heater have if it is to be used on 550 volts and is to take 4 amperes?

$$\text{resistance} = \frac{\text{voltage}}{\text{current}}$$

$$\text{resistance} = \frac{550}{4} = 137.5 \text{ ohms}$$

Example 2. A spotlight takes a current of 11 amperes when used on a 120-volt circuit. What is its resistance?

$$\text{resistance} = \frac{\text{voltage}}{\text{current}}$$

$$\text{resistance} = \frac{120}{11} = 10.9 \text{ ohms}$$

Prob. 1-2. An incandescent lamp uses 0.21 ampere on a 120-volt circuit. What is the resistance of the lamp when burning?

Prob. 2-2. Through what resistance will 230 volts establish 30 amperes?

Prob. 3-2. What is the resistance of the heating element of an electric soldering iron designed to take a current of 2.1 amperes on a 120-volt circuit?

Prob. 4-2. What is the resistance of a vacuum-tube filament which takes 0.3 ampere at 6.3 volts?

Prob. 5-2. The potential across a certain piece of apparatus is found to be 38.7 volts, while the current through it is 3.1 amperes. What is its resistance?

Prob. 6-2. What resistance must an automobile starting motor have if it takes 55 amperes from the car's 12-volt battery when the key is turned?

3. Ohm's Law. Voltage. We may wish at times to find the voltage necessary to set up a certain current in a given resistance. Then we will find that another form of Ohm's law is helpful. If we remember that voltage is proportional to current in a resistance, we can realize that the voltage for a certain current will be higher across a high resistance than the voltage across a lower resistance. Since we *divide* voltage by current to obtain resistance, we must *multiply* resistance by current to obtain voltage. We can now state in words another useful form of Ohm's law:

The voltage required to establish a given current in a given resistance is the product of the resistance times the current.

In briefer form we may write:

$$\text{voltage} = \text{resistance} \times \text{current}$$

or:

$$\text{volts} = \text{ohms} \times \text{amperes}$$

Example 3. The hot resistance of an incandescent lamp is 220 ohms. It requires 0.5 ampere to cause it to glow. What voltage must be impressed across it?

$$\text{voltage} = \text{ohms} \times \text{amperes}$$
$$\text{voltage} = 220 \times 0.5 = 110 \text{ volts}$$

Example 4. To ring a certain bell requires 0.25 ampere. The resistance of the coils in the bell is 12 ohms. What voltage is required?

$$\text{voltage} = \text{ohms} \times \text{amperes}$$
$$\text{voltage} = 12 \times 0.25 = 3 \text{ volts}$$

Prob. 7-2. What voltage will produce a current of 10 amperes in 100 ohms resistance?

Prob. 8-2. What voltage is needed to light the filament of a vacuum tube if its resistance is 42 ohms and its current is 0.3 ampere?

Prob. 9-2. A megohm is 1 million ohms. How many volts are necessary to cause a current of 0.003 ampere in a resistance of 50 megohms?

Prob. 10-2. An electric bell has a resistance of 800 ohms and requires 0.03 ampere to ring. What is the smallest voltage that will ring the bell?

Prob. 11-2. What voltage must be used to operate a telephone receiver having a resistance of 1000 ohms and requiring 0.003 ampere?

Prob. 12-2. A radio pilot light requires 0.055 ampere to make it glow. Its resistance is 100 ohms. What voltage is required?

Prob. 13-2. What voltage exists across a neon lamp of 10,000 ohms resistance when there is a current of 0.02 ampere through it?

4. Ohm's Law. Current. Many practical problems occur in which the current is desired in a given resistance that is subjected to a given volt-

10   Essentials of Electricity

age. Just as there are special forms of Ohm's law for finding resistance or voltage, there is a convenient form for finding current. This form can be stated in words as:

The current that a given voltage can establish in a given resistance equals the voltage divided by the resistance.

This result follows from the fact that, if we *divide* voltage by current to get resistance, we must, in the same way, *divide* voltage by resistance to get current in order to maintain the proportionality of Ohm's law. In briefer form, we may write:

$$\text{current} = \frac{\text{voltage}}{\text{resistance}}$$

or:

$$\text{amperes} = \frac{\text{volts}}{\text{ohms}}$$

Example 5. An electric soldering iron has a resistance (hot) of 70 ohms. What current will it take from a circuit that maintains a potential of 120 volts?

$$\text{amperes} = \frac{\text{volts}}{\text{ohms}} = \frac{120}{70} = 1.71 \text{ amperes}$$

Example 6. What is the current in a hot incandescent lamp, which has a resistance of 27.5 ohms, when used on a 110-volt line?

$$\text{amperes} = \frac{\text{volts}}{\text{ohms}} = \frac{110}{27.5} = 4 \text{ amperes}$$

Prob. 14-2. What current can 100 volts establish in 5 ohms?

Prob. 15-2. How many amperes would be required for an electric iron having 22 ohms resistance if it operates on 110 volts?

Prob. 16-2. An automatic toaster is designed for operation on 120 volts. What current will it take if its resistance is 10.9 ohms?

Prob. 17-2. An electroplating generator creates 6 volts, and the resistance of the circuit is 0.2 ohm. What is the current?

Prob. 18-2. A dry cell has a terminal potential of 1.3 volts when a wire of 0.325 ohm is placed across its terminals. What current exists in the wire?

Prob. 19-2. A battery of dry cells has a terminal potential of 3 volts when placed directly across the filament of a vacuum tube having a resistance of 50 ohms. What is the filament current?

Prob. 20-2. What filament current is taken by a vacuum tube whose filament has a resistance of 4.8 ohms and operates on 1.2 volts?

Prob. 21-2. What will the current be in the coils of an electric bell having a resistance between terminals of 150 ohms when operated on a 6-volt storage battery?

Prob. 22-2. The resistance of a certain tungsten lamp when cold is 20 ohms. What will the current be the instant it is placed across a 110-volt line?

Prob. 23-2. The resistance of the lamp of Prob. 22-2 rises to 84 ohms when glowing at full brilliance. What will be the final steady current of this lamp?

Prob. 24-2. A subway-car heater has a resistance of 100 ohms. What is the current in the heater when the distance of the car from the generating station is such that a potential of 525 volts exists across the heater terminals?

Prob. 25-2. A certain reel of rubber-insulated cable when placed in a tank of conducting liquid has a resistance of 4 billion ohms between the conductor of the cable and a plate immersed in the liquid. What current will there be in the insulation when 300 volts are impressed on this cable?

5. Symbols. When an electric circuit is to be built from many different parts, wiring layouts or even actual photographs of the parts may be necessary to make clear how the parts can be arranged efficiently. However, when the behavior of an electric circuit is to be studied, it is convenient to have some shorthand methods of sketching only what is electrically essential to the layout, eliminating all other details—such details as mounting bolts, thickness of insulation, and necessary length of wires, to list a few. These shorthand symbols are combined in drawings called *circuit diagrams* to show only essential electrical features, the complete layout being left for more complicated manufacturing or construction drawings. A few of the important standard circuit symbols are shown in Fig. 1.

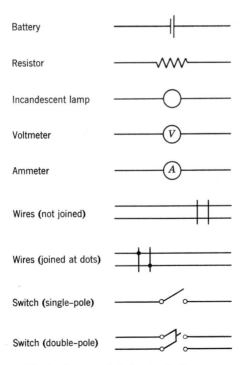

Fig. 1. Some symbols for circuit diagrams.

12   Essentials of Electricity

In Chapter 1 we found that even pieces of wire require some potential whenever a current exists in them. However, in many circuits the wires are chosen large enough in diameter that we can forget their effect on the circuit. For such arrangements the straight lines that we show in a circuit diagram are considered to have no resistance at all, i.e., a current can exist in these ideal wires without requiring any potential difference between their ends. This simplification is extremely important because it directs our attention to the essential elements in the circuit—batteries, lamps, resistors, etc.

Note the symbol for a *resistor*. When we divide voltage by current, we obtain a ratio that we call *resistance*. The actual physical appearance of the section of circuit having resistance depends very much on the total work required by that section. For example, we might find the same resistance ratio for a small element in a radio receiver as we find for a large control element on a diesel-electric locomotive. Therefore, in order to have only one symbol for resistance elements of such different size and shape, we draw the standard broken line, sometimes writing the number of ohms beside the symbol. In other words, an electric element that has the proportional property of *resistance* is called a *resistor*.

6. Electrons. Conventional Current. When Coulomb, an early experimenter, studied electrical effects, he found that charge seemed to be of two kinds. Eventually, men called one positive (+) and the other negative (−). After Volta had made the battery important as an electric source, experimenters came to the conclusion that an electric current is a flow of positive (+) charge. Later work has shown that this conclusion is often incorrect.

Recall that in Chapter 1 we noted that all materials are composed of electric charge. Experimenters have found that the *atom*, or building block, of each material has groupings of positive charge at its center, or *nucleus*. An equal amount of negative charge is arranged in a kind of outer husk around the nucleus. Delicate measurements have shown that this negative (−) charge has a basic building block, called the *electron*. It is now known that when there is an electric current in a *metal*, the only charges that move are some of the electrons from the outer parts of the atoms of the metal. This motion of negative charges in a wire is known as an *electron current*. Unfortunately for calculations on practical devices, the charge of an electron is very small: the charge known as *one coulomb* represents the total charge of 6,240,000,000,000,000,000 electrons.

In apparatus for communication and industrial automation, *electron tubes* are sealed devices in which there is an electron current. In recent

2: Ohm's Law

experiments with the materials known as *semiconductors* (the essential materials for *transistors* in portable radios), men have found that an electric current may be the flow of positive charge, the flow of negative charge, or a combination of both kinds of charge moving in opposite directions. In the face of this complexity, we see that we must specify carefully what we mean by electric current and its direction.

Before the relation of electric charge to the construction of atoms was stated, many practical applications of electricity in motion were made by the early experimenters. They based their work on the idea that an electric current is a flow of positive charge. Although we now understand their work better than they did, we have decided that their specification of positive charge in the flow is extremely useful. Therefore we may state: An electric current is the flow of positive (+) charge.

Because this statement is an agreement (or convention), the flow of positive charge is called *conventional current*. For electron tubes, we may discuss electron current; for transistors, we may wish to discuss both positive-charge current and electron current. However, in this book all problems are discussed from the viewpoint of conventional current. Since this choice is the basis of all the important electric-circuit study today, the name "conventional current" is usually shortened to be simply "current"—the flow of positive (+) charge.

7. Polarity. Current Direction. Although early experimenters left behind some confusion about the *kinds* of charge in motion, they did correctly organize a simple but important rule, called the polarity rule.

Like polarities repel, and opposite polarities attract.

Let us now see how this rule helps us to keep current directions and voltage polarities clear in our thinking.

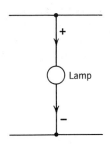

Fig. 2. Arrows indicate conventional current; polarity marks show sense for voltage.

When an electric current (conventional) exists in a certain part of a circuit, we know that work is needed to set up and maintain that current. Suppose we are thinking of current in an incandescent lamp, as in Fig. 2. If the direction of (positive) charge flow in the lamp is found to be downward, we then say that the upper end of the circuit must be plus (+) or positive, while the lower end must be negative. This statement follows from the polarity rule, for the positive upper end must somehow be repelling positive charge down through the lamp, while the lower negative end is attracting positive charge.

14   Essentials of Electricity

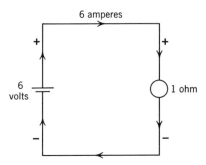

Fig. 3. Polarity relations for a source and its load.

The ideas connected with the lamp in Fig. 2 introduce two new kinds of circuit symbols. The arrows drawn on the wires leading to and from the lamp, or any other electric element, serve to show us the direction of *current* or positive charge flow. The *pair* of plus and minus signs shows us the location and sense of the voltage across the lamp. It is a good idea to think of a *pair* of polarity marks for voltage, so that there will not be confusion if several pairs of signs are marked on a large circuit diagram.

When it is necessary to think of the motion of the individual electrons in the circuit of Fig. 2, the flow of negative charge (electrons) is *upward* through the lamp. The polarity marks for the voltage across the lamp remain in their same locations. In accordance with the polarity rule, the negative lower end of the lamp is repelling electrons upward through the lamp, while the positive upper end is attracting electrons. Note carefully, however, that in this book an arrow shown for a current designates the direction of positive (+) charge motion, i.e., the arrow for a current designates the direction for conventional current.

8. Ideal Battery. In Fig. 1 are shown some of the standard electric-circuit symbols. Although the first one can mean an actual battery, it is also used to mean any source with fixed potential. The value of this potential is usually marked in volts next to the symbol. Recall that ideal wires shown in circuit diagrams are considered to have no resistance. Thus, the wires leading away from the battery symbol are considered to have zero resistance. Consequently, the battery symbol often means an *ideal battery,* i.e., one having fixed voltage and no inherent resistance. This ideal battery may be used in circuit diagrams to mean any fixed voltage, whether it is provided by an actual battery, a rotating generator, or some other source.

Because the ideal battery (or often simply battery) is the source of work that sets up a current, note carefully that current direction is from

2: Ohm's Law   15

minus (−) to plus (+) through the battery while the battery is creating current from plus (+) to minus (−) through lamps or resistors connected to it. Figure 3 shows a lamp, with a hot resistance of 1 ohm, connected to a 6-volt battery. Study the polarity marks carefully to see that the 6-ampere current has a direction from the plus terminal of the battery, around the circuit, and into the negative terminal of the battery. Recall that like polarities repel and unlike polarities attract. Try to fix clearly in mind these polarity and current-direction relations, for they are always found when a battery or other source is doing electrical work and when a device having resistance is absorbing that work.

SUMMARY

OHM'S LAW states the proportional relationship that exists between potential and current, through the constant of proportionality called resistance. It is written in THREE forms:

1. $Ohms = \dfrac{volts}{amperes}$, or, $resistance = \dfrac{voltage}{current}$

2. $Volts = ohms \times amperes$, or, $voltage = resistance \times current$

3. $Amperes = \dfrac{volts}{ohms}$, or, $current = \dfrac{voltage}{resistance}$

SYMBOLS used in CIRCUIT DIAGRAMS are the shorthand for writing the essential features of an electric arrangement. WIRES of a circuit diagram merely show connections and have no resistance at all.

DIRECTION of current or positive charge flow is from plus to minus in a resistor.

IDEAL BATTERY means a source having fixed potential and no inherent resistance.

PROBLEMS

Prob. 26-2. If a subway-car heater is supplied with a potential of 550 volts from the third rail, what must the heater's resistance be to limit the current to 10 amperes?

Prob. 27-2. A d-c generator, G, shown connected as in Fig. 4, is producing a potential of 127.5 volts at its terminals. How much current exists when the resistance connected across its terminals is 63 ohms?

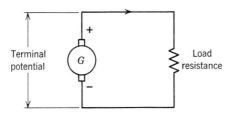

Fig. 4. Generator G is the source.

16 Essentials of Electricity

Prob. 28-2. What potential must an electronic source produce to create an electroplating current of 100 amperes in a circuit whose total resistance is 0.25 ohm?

Prob. 29-2. In a connection like that of Fig. 4, a d-c generator has a terminal potential of 35 volts when the external load resistance is 0.85 ohm. What is the magnitude of the load current?

Prob. 30-2. In a connection like that of Fig. 4, the load resistance has a potential across it of 150 volts. What must the size of the load resistance be, if the current between it and the generator is 5 amperes?

Prob. 31-2. A lamp for a 220-volt circuit has a resistance of 480 ohms. What will its current be?

Prob. 32-2. The field winding of an electric motor has a resistance of 55 ohms. What is the field current when the potential across the winding is 232 volts?

Prob. 33-2. A crane motor has a field resistance of 0.021 ohm. What is the voltage across the field winding when the current is 76 amperes?

Prob. 34-2. The rotor winding of a d-c generator has a resistance of 0.033 ohm. What will the voltage across this resistance be when there is a current of 115 amperes in it?

Prob. 35-2. Which resistance is the greatest of these three: No. 1, with 5.19 amperes at 12 volts; No. 2, with 263 amperes at 550 volts; or No. 3, with 0.0035 ampere at 0.0015 volt?

Prob. 36-2. Under certain conditions of contact, the resistance of the human body is about 10,000 ohms from hand to hand. What current would exist in the arms if the potential applied between the hands were 6600 volts?

Prob. 37-2. An electric oven requires 21 amperes at 440 volts when operated at full heat. What is its resistance under these conditions?

Prob. 38-2. The resistance of a restaurant's cooking apparatus becomes 2.4 ohms when all units are turned on. What current will exist in the supply feeders if the terminal potential at the equipment is 240 volts?

Prob. 39-2. The accompanying table shows data for measurements on a certain tungsten lamp. Plot a curve with volts measured horizontally and amperes vertically, thus graphically showing the relation between current and terminal potential for this lamp. Are potential and current proportional for this lamp?

Compute the resistance for each of the potentials and plot another curve using resistance instead of current on the vertical scale. The cold resistance is 131 ohms. Is the resistance (proportionality ratio) constant for this device?

VOLTS	AMPERES	VOLTS	AMPERES
6	0.03	80	0.158
10	0.044	110	0.192
20	0.066	160	0.235
30	0.085	210	0.276
50	0.118		

3

Simple Electric Circuits

1. Definition of Series and Parallel Circuits. There are two important ways of connecting two or more pieces of electrical apparatus.

Series. When the pieces are connected one after the other, they are said to be in *series*. Lamps *A* and *B* of Fig. 5 are in series.

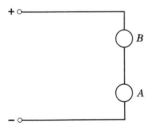 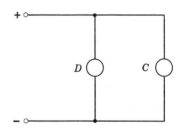

Fig. 5. Series connection.　　　　　　Fig. 6. Parallel connection.

Parallel. When the pieces are connected side by side so that the current is divided between them, they are said to be in *parallel* with one another. *Multiple* or *shunt* are other names sometimes used for this parallel connection. Lamps *C* and *D* of Fig. 6 are in parallel with each other.

The two combinations may exist in the same circuit. In Fig. 7, the parallel combination of resistors *C* and *D* is in series with the series

18 Essentials of Electricity

combination of resistors *A* and *B*. Also, as shown in Fig. 8, the parallel combination of lamps *C* and *D* is in parallel with the series combination of resistors *A* and *B*. However, in this chapter only simple series and simple parallel circuits are considered.

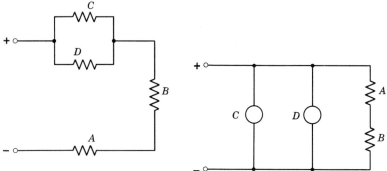

Fig. 7. Series connection of a parallel combination.

Fig. 8. Parallel connection of a series grouping.

2. Series Circuit. Current. If a series of pipes of unequal diameters are joined together, end to end, there can be only one rate at which water flows through them, because the water cannot escape into any other path. Similarly, in a series electric circuit, the average rate at which charge passes any point of that circuit must be the same everywhere, because the charge does not escape from the circuit.

Suppose we join together in series several electrical devices. In the series circuit of Fig. 9 are a 10-ohm heater resistor (R), an electric bell (B) with a resistance of 50 ohms, and an incandescent lamp (L) with a resistance of 200 ohms. Although the resistance of the heater is much lower than that of the bell or the lamp, no greater current can exist in the heater than in the bell or lamp. The electric current at one end of a series connection must be the same as the current at the other end, just as the rate at which water entering a series of pipes must be the same as the rate at which water leaves the series, if there are no leaks.

The first fact to be noted, then, is that:

In a series circuit the current is the same in all parts, no matter what the resistance of each part may be.

3. Series Circuit. Potential. Suppose that a resistor (R) having 4 ohms resistance is joined in series with a lamp (L) having 16 ohms resistance, as in Fig. 10. When this combination is connected to a certain source, the current in the series circuit is 6 amperes. Let us determine the

3: Simple Electric Circuits   19

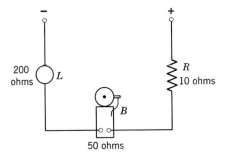

Fig. 9. The current is the same in each element of a series connection.

potentials of this circuit by Ohm's law. Points *a* through *f* have been marked in Fig. 10 to help us identify parts of this circuit easily.

In Section 2-3 we found that the voltage required to establish a given current in a given resistance is the product of the resistance times the current. Therefore, the voltage between points *b* and *c* (the voltage across resistor *R*) is:

$$\text{volts} = \text{ohms} \times \text{amperes}$$
$$= 4 \times 6 = 24 \text{ volts}$$

In other words, there must be a potential of 24 volts between points *b* and *c* while there is a current of 6 amperes in the 4-ohm resistor. Furthermore, point *b* must be positive with respect to point *c*, in accordance with the ideas of polarity set forth in Section 2-7. Thus, we would mark the diagram with a *pair* of signs, plus (+) at *b*, and minus (−) at *c*. In a similar fashion, the voltage across the lamp is

$$\text{volts} = \text{ohms} \times \text{amperes}$$
$$= 16 \times 6 = 96 \text{ volts}$$

We would mark the diagram (+) at *d* and (−) at *e* to show the sense of this potential of 96 volts between *d* and *e*, or across the lamp.

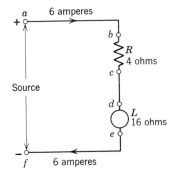

Fig. 10. Resistor and lamp in series.

Now we should recall that potential, as in Section 1-4, refers to work. In order to establish 6 amperes in resistor *R*, a potential of 24 volts is required, or 24 joules of work must be done if 1 coulomb of charge is transported between points *b* and *c*. Similarly, 96 joules of work must be done if 1 coulomb were transported at the rate of 6 amperes between points d and e. Because the circuit of Fig. 10 is a series circuit, the current is the same in all parts. If 24 joules of work are required for the passage through resistor R, 96 *more* joules of work must be done to maintain the same 6-ampere current through lamp L. Therefore, the total work for 1 coulomb, or the total potential to establish 6 amperes between b and e is

$$24 + 96 = 120 \text{ volts}$$

Because the current is the same in all parts of a series circuit, the total work done in establishing the current must be the sum of all the individual works needed to set up that current in each of the parts. Thus it is always true in a series circuit that, if we *add* up the potentials across all the parts in series, the sum will be the potential across the series combination.

The second fact to be noted, then, is that:

The voltage across the parts in series equals the sum of the voltages across the separate parts.

4. Series Circuit. Resistance. It is helpful to recall that the ideal wires connecting the symbols in the circuit diagrams have no resistance themselves. Thus, in Fig. 10, the wire between c and d has zero resistance, zero resistance meaning that the 6-ampere current through the wire requires no voltage between c and d. Therefore, we say that c and d are at the same potential. By the same reasoning we say that points a and b are at the same voltage, although the potential of a and b is 24 volts more than the potential of c and d. Also, points f and e share a common potential, one that is different from either that at a and b or that at c and d. Figure 11 shows the circuit of Fig. 10 with all the polarities and potentials added.

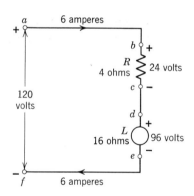

Fig. 11. The voltage across the parts in series equals the sum of the voltages across the separate parts.

In Section 2-2 we saw that resistance equals the quotient of voltage divided by current. When we apply this idea to the source terminals of Fig. 11, we have

$$\text{resistance} = \frac{\text{voltage}}{\text{current}}$$

$$= \frac{120}{6} = 20 \text{ ohms}$$

At the terminals of the source, this series circuit behaves as though it were a resistance of 20 ohms. We can make this number have another meaning if we add together the resistance of the separate parts of the series circuit of Fig. 11. The total resistance equals the resistance of lamp L plus the resistance of resistor R, or

$$16 + 4 = 20 \text{ ohms}$$

Since, as noted above, the ideal connecting wires of our symbols do not have any resistance, we need add together only the resistances of the parts that we have called to our attention with the special electrical symbols for lamps, resistors, etc.

Therefore, the third fact to be noted about a series circuit is that:

The total resistance of parts in series is the sum of the separate resistances.

Example 1. (*a*) What is the resistance of the circuit of Fig. 12? (*b*) What is the current in this circuit?

$$\text{resistance} = 200 + 200 + 40 = 440 \text{ ohms}$$

$$\text{current} = \frac{\text{volts}}{\text{ohms}} = \frac{220}{440} = 0.5 \text{ ampere}$$

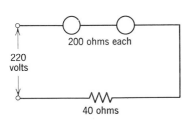

Fig. 12. The two lamps and resistor are in series across a 220-volt line.

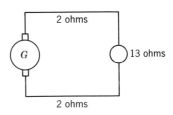

Fig. 13. The resistive line wires form a series connection with the lamp.

Prob. 1-3. (*a*) What is the total resistance of the circuit connected across the terminals of the generator in Fig. 13? (The connecting wires have definite resistance here.)

(*b*) What current exists in the circuit while the generator maintains a terminal potential of 110 volts?

Prob. 2-3. (*a*) If 10 lamps of 13 ohms each are connected in series across the generator in place of the single lamp of Problem 1-3, what will the total resistance be?

(*b*) What will the current be under these conditions?

(*c*) If the resistance is assumed to increase to 1.2 times as much as it was in (*b*) when the current doubles, what voltage will be required from the generator to double the current through the ten lamps?

Prob. 3-3. What resistance must each lamp of Fig. 14 have if the current in the circuit is to be limited to 6.6 amperes? Neglect the resistance of the line wires.

22 Essentials of Electricity

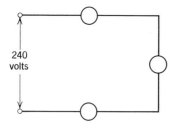

Fig. 14. Three airport landing lights in series.

5. Series Circuit. Current, Voltage, and Resistance. The three facts which should be learned with regard to a series circuit may be tabulated as in Fig. 15.

Example 2. A current of 3 amperes exists in the series circuit of Fig. 16.
(*a*) What is the potential across *AB*?
(*b*) What is the potential across each part of the circuit?

Series Combination

Current in a series combination is	*same* as current in each separate part.
Voltage across a series combination is	*sum* of the voltage across the separate parts.
Resistance of a series combination is	*sum* of the resistances of the separate parts.

Fig. 15. Summary of series-circuit relationships.

The total resistance = 10 + 40 + 15 = 65 ohms. The potential to create 3 amperes in 65 ohms = 65 × 3 = 195 volts. Answer to (*a*): Potential across *AB* is 195 volts.

(*b*) Potential needed for 3 amperes in 10-ohm resistor = 10 × 3 = 30 volts.
Potential needed for 3 amperes in 40-ohm lamp = 40 × 3 = 120 volts.
Potential needed for 3 amperes in 15-ohm resistor = 3 × 15 = 45 volts.
Potential needed for 3 amperes in the complete circuit = 30 + 120 + 45 = 195 volts.

This last answer checks with that of part (*a*), found by multiplying the total resistance by the current.

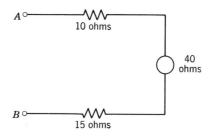

Fig. 16. Two resistors and a lamp in series.

Prob. 4-3. (*a*) If each lamp of Fig. 17 is to have a current of 1.7 amperes in it, what must be the current in the generator?

(*b*) What is the total resistance of the lamps connected across the terminals of the generator if the separate resistances are: $A = 67$ ohms, $B = 150$ ohms, $C = 200$ ohms?

(*c*) When the current is 1.7 amperes in each lamp, what is the terminal potential of the generator?

Prob. 5-3. An airport runway lamp requires 6.6 amperes to provide sufficient illumination. If the resistance is 15 ohms for that illumination, what voltage must a generator supply to seven such lamps in series?

Prob. 6-3. In Fig. 18 motor M, supplied from generator G, requires a current of 50 amperes with a terminal potential of 230 volts. Each line wire has a resistance of 0.25 ohm. What terminal potential must generator G supply?

6. Application of Ohm's Law. It should be noted that in solving Example 2, with Fig. 16 Ohm's law was used in (*a*) to find the voltage necessary to establish the current in the whole circuit of the three resistances.

First, the total resistance was found by adding $10 + 40 + 15 = 65$ ohms. Then, the total voltage necessary was:

(total) voltage = (total) current × (total) resistance
(total) voltage = $3 \times 65 = 195$ volts

Note that to find the total voltage, it was necessary to use the *total* current and *total* resistance.

In (*b*) of Example 2, when we wished to find the voltage necessary to establish the current in the 10-ohm resistor *only,* we used Ohm's law again. But this time, since we wanted *only* the voltage across the 10-ohm resistor, we used *only* the resistance of the 10 ohms, not the total resistance of the circuit. We also had to use *only* the current through the 10-ohm resistor. Thus we said:

voltage (across 10-ohm resistor) = resistance (of 10-ohm resistor ×
current (in 10-ohm resistor) = $10 \times 3 = 30$ volts

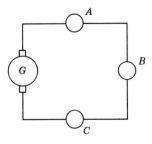

Fig. 17. Lamps in series across the terminals of a generator.

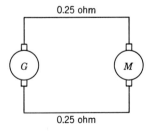

Fig. 18. Generator G supplies the series connection of motor M and the resistive line wires.

24   Essentials of Electricity

To find the voltage across the *10-ohm resistor,* we used the resistance of the *10-ohm resistor* and the current in the *10-ohm resistor.*

Recall that Ohm's law is a law of proportionality. If a portion of a circuit has a certain resistance, that resistance, or proportionality ratio, expresses the relation between the sizes of the voltage across that portion and the current in that portion. If we look at the total resistance of a series string of elements, we are thinking of another proportionality ratio, which expresses the relation between the sizes of the voltage across the entire circuit and the current that is common in each of the elements. Hence we see that Ohm's law may be applied either to the whole of a circuit or to any part of a circuit.

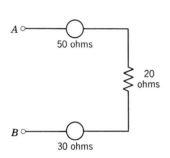

Fig. 19. Two lamps and resistor in series.

But if Ohm's law is applied to the *whole* circuit, the voltage, resistance, and current must be the voltage, resistance, and current of the *whole* circuit, and not of merely a part. And if it is applied to a *part* of a circuit, the voltage, current, and resistance must be the voltage, current, and resistance of *that part only.* This principle is vitally important in applying Ohm's law. Many mistakes are made in the use of this simple law of proportionality, just because there is a failure to work with the voltage current and resistance of the *same* part of the circuit. Study of Example 3 will show the correct application of Ohm's law.

Example 3. In the circuit of Fig. 19 a lamp with resistance of 50 ohms, a resistor with resistance of 20 ohms, and another lamp with resistance of 30 ohms are connected in series across the points A and B. The 30-ohm lamp is found to have a potential of 120 volts across it.

Find:

(*a*) Current in each part.
(*b*) Voltage across each part.
(*c*) Voltage across AB.

We are able to find the current in the 30-ohm lamp, because we know *both* the resistance and the potential.

$$\text{current (through 30-ohm lamp)} = \frac{\text{potential (across 30-ohm lamp)}}{\text{resistance (of 30-ohm lamp)}}$$

$$= \frac{120}{30} = 4 \text{ amperes}$$

Thus the current in the 30-ohm lamp is 4 amperes. But since the 50-ohm lamp and the 20-ohm resistor are in series with the 30-ohm lamp, the current must be the same in all of them. Therefore, we can find the voltage across the 20-ohm resistor:

voltage (across 20-ohm resistor) = resistance (of 20-ohm resistor)
× current (in 20-ohm resistor)
= 20 × 4 = 80 volts

voltage (across 50-ohm lamp) = resistance (of 50-ohm lamp)
× current (in 50-ohm lamp)
= 50 × 4 = 200 volts

Since this circuit is a series connection, the potential across the whole circuit, that is, across AB, equals the sum of the potentials across the separate parts, or

potential across AB = 120 + 80 + 200 = 400 volts.

The answers are, then,

(a) current in each in each part = 4 amperes
(b) voltage across 20-ohm resistor = 80 volts
 voltage across 50-ohm lamp = 200 volts
(c) voltage across AB = 400 volts

Note that after we had found the current of the *whole* circuit, we might have added up the resistances and found the resistance of the *whole* circuit, thus 30 + 20 + 50 = 100 ohms. Then potential (across total circuit) = resistance (of total circuit) × current (in total circuit).

100 × 4 = 400 volts

Of course this answer checks with the potential found by the first method, for either method is an application of Ohm's law of proportionality.

Prob. 7-3. Lamp L, of Fig. 20, requires 7 amperes. The generator terminal potential is 220 volts. Each line wire has a resistance of 2 ohms.

(a) What is the voltage drop (the potential required) in the line wires?

(b) What is the voltage at the lamp terminals?

(c) What is the resistance of the lamp?

Prob. 8-3. A 10-ohm resistor is connected in series with a 15-ohm resistor.

(a) What voltage must be placed across this combination to establish a current of 5 amperes in it?

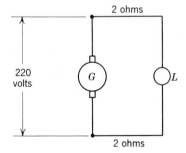

Fig. 20. Lamp L is in series with resistive line wires.

(b) Under the conditions of (a), what would the voltage be across each of the resistors?

(c) For what voltage across the combination would the voltage across the 15-ohm resistor be 60 volts?

26   Essentials of Electricity

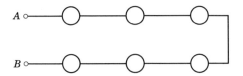

Fig. 21. Series-connected street lamps.

Prob. 9-3. Six street lamps are joined in series by seven equal lengths of wire having a total resistance of 10 ohms, as shown in Fig. 21. Each lamp has a resistance of 14 ohms. What voltage is required between points A and B to set up a current of 6.6 amperes in the circuit?

Prob. 10-3. (*a*) What is the potential across each lamp of Problem 9-3?

(*b*) If one lamp should become short-circuited, what new voltage would be required between points A and B to maintain the current at 6.6 amperes?

Prob. 11-3. In an arc welder, the current in the arc is 200 amperes. What is the resistance of the arc stream when the voltage across it is 50 volts?

7. Parallel Circuit. Voltage. Suppose that water is to be taken from a reservoir high up in a mountainous region down to another reservoir near a city. Suppose that three pipes laid side by side carry the water down. Even if each of the three pipes has a different diameter, the change in level of water from the upper to the lower reservoirs is the same for each of the three pipes. Since the difference in water level determines the pressure forcing current through the pipes, each pipe has the same pressure across it. Water enters each pipe at the same high pressure, and the water leaves each pipe at the same low pressure. For each pipe the pressure is the difference in level between the two reservoir levels. If the difference in level of the two reservoirs does not change, many more pipes may be laid in parallel between the high and the low levels, and the pressure across all the pipes will still be the same.

Similarly, suppose we join three electrical devices in parallel, as shown in Fig. 22, between two distribution wires R and S, between which the potential level is kept fixed at 120 volts. Since all three elements A, B,

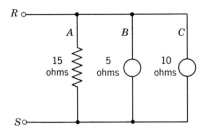

Fig. 22. Branches A, B, and C are connected in parallel.

and C lie between R and S, the potential difference between R and S is also the potential or voltage across each element. Therefore, each element has 120 volts across it. The voltage across resistor A is 120 volts, the voltage across lamp B is 120 volts, and the voltage across lamp C is 120 volts.

The first fact to be noted for a parallel circuit is that:

The voltage across the parallel combination is the same as the voltage across each branch.

8. Parallel Circuit. Current. Think again about the three pipes between the two reservoirs. Since each of the pipes is independent of the others, each pipe can have an amount of water flowing through it that is different from the amount flowing in the others. In fact, the pipe with the largest diameter will carry the largest flow of water, and the pipe with the smallest diameter will have the smallest flow. It is not true that in these three paths the water takes *only* the path of least resistance. Water flows down from the upper to the lower reservoir through all three pipes, but with different rates of flow through each pipe. The total flow of water is the sum of the three separate amounts flowing in the three pipes.

Similar reasoning applies to the parallel circuit. Each branch of the circuit of Fig. 22 has the same difference of potential across it: 120 volts. Since the opposition to setting up a current is highest for branch A because branch A has the highest resistance, the smallest current will be found in branch A. By applying Ohm's law to this branch alone, we can determine the size of its current. Since we know that branch A has 120 volts across it and that the resistance of branch A is 15 ohms, we can apply Ohm's law as follows:

$$\text{current (through } A) = \frac{\text{voltage (across } A)}{\text{resistance (of } A)}$$
$$= \frac{120}{15} = 8 \text{ amperes}$$

The other two currents may be found in the same way.

$$\text{current (through } B) = \frac{\text{voltage (across } B)}{\text{resistance (of } B)}$$
$$= \frac{120}{5} = 24 \text{ amperes}$$

$$\text{current (through } C) = \frac{\text{voltage (across } C)}{\text{resistance (of } C)}$$
$$= \frac{120}{10} = 12 \text{ amperes}$$

28 Essentials of Electricity

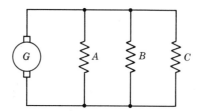

Fig. 23. The same voltage exists across each of the branches A, B, and C.

The current entering and leaving in the distribution wires R and S is then the sum of these three branch currents, or $8 + 24 + 12 = 44$ amperes. Since Ohm's law has been satisfied separately for each branch, and since each branch receives its current from wire R and each branch returns its current to wire S, the total current in wires R and S must be the sum of the branch currents of the parallel branches, as added up above.

The second fact to be noted for a parallel circuit is that:

The total current into the parallel combination is the sum of the currents in the separate branches or paths.

Prob. 12-3. Generator G of Fig. 23 maintains a potential of 12 volts across resistors A, B, and C of 2, 5, and 14 ohms resistance, respectively. What is the current in each, and what is the current supplied by the generator?

Prob. 13-3. Each of the six paralleled lamps in Fig. 24 requires 0.55 ampere. How much current is supplied by the generator?

9. Parallel Circuit. Resistance. Suppose that we would like to find the resistance of a parallel combination, as A, B, and C in Fig. 22. An easy method of calculation is to apply Ohm's law in several steps as follows:

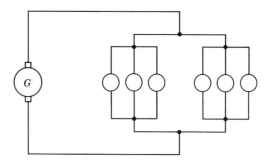

Fig. 24. The group of six lamps in parallel requires 6 times the current of one lamp.

Previous calculation gave:

$$\text{current through } A = \frac{120}{15} = 8 \text{ amperes}$$

$$\text{current through } B = \frac{120}{5} = 24 \text{ amperes}$$

$$\text{current through } C = \frac{120}{10} = 12 \text{ amperes}$$

current through combination = 44 amperes

Now since we know the current through the *combination* (44 amperes) and the voltage across the *combination* (120 volts), we can find the resistance of the combination.

$$\text{resistance (of combination)} = \frac{\text{voltage (across combination)}}{\text{current (through combination)}}$$

$$= \frac{120}{44} = 2.73 \text{ ohms}$$

At first sight it may seem strange that the resistance of a combination of three circuit elements of 5, 10, and 15 ohms should be only 2.73 ohms. But the apparent difficulty disappears when we consider that the more paths we have in parallel in which there are currents from one point to another, the lower the resistance is between those two points. Thus, if there had been only the 5-ohm path between the points R and S, the resistance between these points would have been 5 ohms. But when another path with a resistance of 10 ohms was connected between the same points R and S, the total current between R and S increases. Since the potential was still 120 volts, the resistance between the two points must be smaller than 5 ohms if the current is to be larger than before. When the third resistance of 15 ohms was added in parallel, the total current became still larger. Because the current is larger between points R and S while the potential remains 120 volts; the resistance or opposition to current must be smaller with the three paths connected than with only two. Thus we may see that the resistance of any parallel combination is less than the resistance of the path of smallest resistance. The path having the smallest resistance in Fig. 22 is the 5-ohm branch, and the combined resistance of the three parallel branches is only 2.73 ohms.

Suppose we were given the three parallel resistances A, B, and C of Fig. 25, with no mention of the voltage across them. We could find the resistance of the parallel combination as follows:

First find the current set up by assuming one volt across each branch.

current (per volt) through $A = \frac{1}{15} = 0.0667$ ampere
current (per volt) through $B = \frac{1}{5} = 0.2$ ampere
current (per volt) through $C = \frac{1}{10} = 0.1$ ampere
current through combination = sum = 0.3667 ampere

If one volt sets up a current of 0.3667 ampere through the combination,

$$\text{resistance (of combination)} = \frac{\text{voltage (across combination)}}{\text{current (through combination)}}$$

$$= \frac{1}{0.3667} = 2.73 \text{ ohms}$$

This result checks with the value found above.

In the preceding calculations we have used the expression "current per volt," as for instance:

current per volt through $A = \frac{1}{15}$ ampere
current per volt through $B = \frac{1}{5}$ ampere

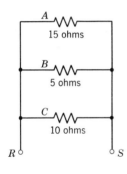

Fig. 25. The equivalent resistance between terminals R and S is 2.73 ohms.

Note that the value $\frac{1}{15}$ is the reciprocal (or inverse) of the resistance of branch A (15 ohms), and the value $\frac{1}{5}$ is the reciprocal of 5, the resistance of branch B (5 ohms). The reciprocal of the resistance of a branch is sometimes called the *conductance* of the branch. Whereas the resistance is a measure of the opposition that a branch offers to a certain voltage in setting up a current, the conductance is a measure of how much current that branch can conduct for the given applied voltage. In order to show the reciprocal relation, it is customary to express the size of a conductance in terms of a unit called the *mho* (ohm spelled backward). Thus, for Fig. 25:

15 ohms resistance has $\frac{1}{15}$ mho conductance or 0.0667 mho
5 ohms resistance has $\frac{1}{5}$ mho conductance or 0.2 mho
10 ohms resistance has $\frac{1}{10}$ mho conductance or 0.1 mho

Note that the larger the resistance is, the smaller the conductance becomes. The conductance of a parallel combination of resistive elements is the sum of the conductances of the separate branches. Thus

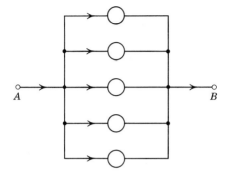

Fig. 26. The resistance between A and B is only $\frac{1}{5}$ that of one lamp.

0.3667 mho is the conductance of the parallel combination of Fig. 25. Conductance, or reciprocal resistance, is merely another name for the term "amperes per volt," or "current per volt."

When the separate branches of a parallel combination have the same resistance, finding the equivalent resistance of the combination is simpler. Suppose that in Fig. 26 the resistance of each lamp is 200 ohms. Since the current supplied from A to B has 5 paths of equal opposition in parallel, the total opposition must be only $\frac{1}{5}$ as great as the opposition or resistance of one path. Thus the resistance of the parallel combination equals $\frac{200}{5} = 40$ ohms.

If we were told that the resistance of the combination was 40 ohms and if we were asked to find the resistance of each lamp, it would seem rather strange to multiply the resistance of the combination (40 ohms) by 5 to get the resistance of one path. But this method would be correct because the resistance of only one path must be five times as much as the resistance of five parallel paths.

The third fact to be noted about a parallel circuit is that:

The equivalent resistance of a parallel combination may be found by successive applications of Ohm's law.

In review, the outline of the calculation for parallel circuits follows. First find the current in each branch. Add these to find the total current into the combination. Then the equivalent resistance of the combination equals the voltage across the combination divided by the current through the combination. If no voltage is given, use 1 volt. The equivalent resistance may also be found by adding together the separate branch conductances in mhos; then the equivalent resistance is the reciprocal of the total conductance.

Example 4. Resistance of 2 ohms, 3 ohms, and 4 ohms are joined in parallel. What is the resistance of the combination?

current (per volt) through 2 ohms $= \frac{1}{2} = 0.5$ amp
current (per volt) through 3 ohms $= \frac{1}{3} = 0.333$ amp
current (per volt) through 4 ohms $= \frac{1}{4} = 0.25$ amp
current (per volt) through combination $=$ sum $= 1.083$ amp

$$\text{resistance (of combination)} = \frac{\text{voltage (across combination)}}{\text{current (through combination)}}$$

$$= \frac{1}{1.083} = 0.923 \text{ ohm}$$

Calculation with conductances:

conductance of 2-ohm branch $= \frac{1}{2} = 0.5$ mho
conductance of 3-ohm branch $= \frac{1}{3} = 0.333$ mho
conductance of 4-ohm branch $= \frac{1}{4} = 0.25$ mho
conductance of combination $=$ sum $= 1.083$ mhos

Therefore, the equivalent resistance of the parallel combination is:

$$\text{resistance} = \frac{1}{\text{conductance}} = \frac{1}{1.083} = 0.923 \text{ ohm}$$

The three facts which should be known about a parallel combination of branches in an electric circuit may be tabulated in the form shown in Fig. 27. These statements should be compared with the corresponding ones for a series circuit, as tabulated in Fig. 15.

Parallel Combination

Current through parallel combination equals	*sum* of currents through separate branches.
Voltage across parallel combination is	*same* as voltage across each branch.
Resistance of a parallel combination is	*less than* resistance of the branch having smallest resistance and may be found by successive use of Ohm's law.

Fig. 27. Summary of parallel-circuit relationships.

Prob. 14-3. A parallel circuit has branches with respective resistances of 1, 3, 5, 10, and 20 ohms.
(*a*) What is the conductance of each branch?
(*b*) What is the conductance of the combination?
(*c*) What is the resistance of the combination?

Prob. 15-3. The cathode heaters of several vacuum tubes are often operated

in parallel across a common source of potential. If 7 tubes, each rated at 12.6 volts, 0.15 ampere, are to be operated in parallel,
 (*a*) What is the resistance of each heater?
 (*b*) What is the resistance of the seven in parallel?
 (*c*) What will be the current and voltage required by the combination?

Prob. 16-3. The lamps and other appliances attached to house-lighting circuits are usually connected in parallel. When No. 14 wire is used for each branch circuit, a safe value of current for that wire is 15 amperes. How many 0.85-ampere lamps can safely be lighted at one time on one of these branch circuits? How many of these lamps could be safely used when a 10-ampere electric toaster is operating from the same branch?

Prob. 17-3. Lamps having resistances of 121 ohms, 242 ohms, and 807 ohms are operated in parallel from the same source. If the current in the 121-ohm lamp is 0.91 amperes, what are the currents in each of the other lamps, and what is the current supplied by the source?

Prob. 18-3. When four identical lamps operate in parallel from the same 110-volt source, the source supplies a current of 3.55 amperes. What is the resistance of each lamp?

Prob. 19-3. When an automobile is started while the headlights are on, the ignition key momentarily connects the starting motor in parallel with the lamps and the battery. Under these conditions, the terminal potential of the battery becomes 3.5 volts, and the starting motor draws a current of 200 amperes. Ordinarily the lamps require a total current of 15 amperes when the battery terminal potential is 6 volts. With the assumption that the resistance of the lamps does not change when the voltage drops, calculate the total current supplied by the battery at the instant the motor is started.

SUMMARY

Electric-circuit elements connected ONE AFTER THE OTHER are said to be in SERIES.

Electric-circuit elements connected SIDE BY SIDE are said to be in PARALLEL.

For review, Fig. 15 is repeated here; see also Fig. 27 on page 32.

Series Combination

Current in a series combination is	*same* as current in each separate part.
Voltage across a series combination is	*sum* of the voltage across the separate parts.
Resistance of a series combination is	*sum* of the resistances of the separate parts.

Fig. 15. Summary of series-circuit relationships.

OHM'S LAW applied to any electric circuit should read:

The amperes through any PART of a circuit equal the volts across that same PART of the circuit divided by the ohms of that same PART of the circuit.

Essentials of Electricity

PROBLEMS

Prob. 20-3. As shown in Fig. 28, a generator supplies current to a parallel combination of three lamps: A, B, and C.

resistance of $A = 60$ ohms
resistance of $B = 40$ ohms
resistance of $C = 90$ ohms

The generator terminal potential is 120 volts. Determine:
(a) Voltage across each lamp.
(b) Current through each lamp.
(c) Current through the combination.
(d) Resistance of the combination.

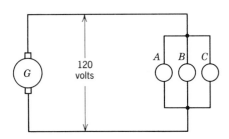

Fig. 28. Three lamps in parallel.

Prob. 21-3. If each lamp of the parallel circuit of Fig. 28 were replaced by one requiring a current of 0.85 ampere, how many amperes must the generator supply to this new load?

Prob. 22-3. In Prob. 20-3, what is the conductance of each lamp? What is the conductance of A and B as a parallel combination? What is the conductance of A, B, and C as a parallel combination?

Prob. 23-3. At a point in a circuit, current divides among three parallel branches having resistances of 5, 10, and 20 ohms, respectively. What single value of conductance could replace this combination of these three branches in the circuit? What is the equivalent resistance of the parallel combination?

Prob. 24-3. There is a current of 20 amperes in the 5-ohm branch of Prob. 23-3. Find the current in each of the other branches and the current required by the parallel combination.

Prob. 25-3. In the circuit of Fig. 29, the two branch resistances are $X = 0.30$ ohm and $Y = 0.20$ ohm. What is the conductance of each branch? What is the total conductance? What is the resistance of the combination?

Prob. 26-3. Find the equivalent resistance of the parallel combination of the two branches in Fig. 29 if $X = 3.00$ ohms and $Y = 2.00$ ohms.

Prob. 27-3. Answer the questions of Prob. 25-3 when resistors X and Y are each equal to 0.30 ohm.

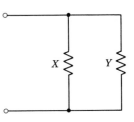

Fig. 29. Two resistors in parallel.

Prob. 28-3. If fourteen lamps of 226 ohms each are connected in parallel across a potential of 104 volts, find:
(a) Resistance of the combination.
(b) Current through each lamp.
(c) Current through the combination.

Prob. 29-3. The generator of Fig. 30 supplies a current of 7.5 amperes, and the group of lamps has a potential of 104 volts across the terminals. Determine:
(a) Current through each lamp.
(b) Resistance of each lamp.
(c) Voltage across each lamp.
(d) Equivalent resistance of the combination.

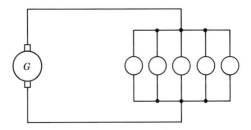

Fig. 30. The generator supplies five lamps in parallel.

Prob. 30-3. If one lamp in Fig. 30 becomes open-circuited, to what potential must the generator be adjusted if the current that the generator supplies is to remain 7.5 amperes? Consider the resistance of each of the remaining four lamps to stay the same as in Prob. 29-3.

Prob. 31-3. Three resistors are connected in parallel. The resistances are: 12 ohms, 21 ohms, and the third resistance unknown. The resistance of the parallel combination is 4.85 ohms. What is the resistance of the third resistor?

Prob. 32-3. If the 12-ohm resistor of Prob. 31-3 has a current in it of 2.3 amperes, what is the current in each of the other branches and in the combination?

Prob. 33-3. What potential is required to set up a current of 12 amperes through a parallel combination of three branches having respective resistances of 15.3 ohms, 1.3 ohms, and 10.5 ohms? What will be the current in each branch?

Prob. 34-3. In the circuit of Fig. 31, lamp B has a resistance of 18 ohms; lamp A has a resistance of 84 ohms and a current of 1.32 amperes. The line wire supplying the circuit has a current of 11 amperes. Determine:
(a) Voltage across the parallel combination.
(b) Current through heater resistor R.
(c) Resistance of R.
(d) Resistance of the entire combination.

Prob. 35-3. In a battery-operated portable radio, the heater elements for four vacuum tubes are operated in parallel. Three of the heaters require 0.06 ampere each, and the other requires 0.12 ampere. What is the total current taken from the battery?

Prob. 36-3. The potential between the third rail and the running rails of a

36 Essentials of Electricity

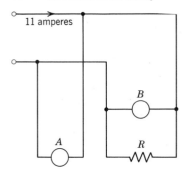

Fig. 31. The parallel combination of lamp B and heater R is in parallel with lamp A.

subway is 550 volts. In order to operate a 110-volt 0.25-ampere lamp from this potential, how much resistance must be placed in series with the lamp?

Prob. 37-3. If five of the lamps described in Prob. 36-3 are connected in series across the 550-volt subway supply, what will the current be through the combination? What will be the current in each lamp? How much voltage will each lamp have across it?

Prob. 38-3. Five lamps, supposedly similar, actually have hot resistances of 484, 403, 512, 475, and 475 ohms, respectively. If these lamps are connected in series across a 550-volt subway supply,

(a) What will be the current in each lamp?
(b) What will be the potential across each lamp?
(c) Which lamp would glow brightest of the five?

Prob. 39-3. An airport runway lamp has a resistance of 8 ohms and requires a current of 6.6 amperes for operation in fog. How many of these lamps can be operated in series on a 1000-volt line? Neglect the resistance of the connecting wire.

Prob. 40-3. At the terminals of a generator the potential is found to be 110 volts when a wire having a resistance of 0.02 ohm is connected across the terminals. What is the current in the wire?

Prob. 41-3. What total current would the generator of Prob. 40-3 supply if an incandescent lamp having a hot resistance of 484 ohms were connected in parallel with the wire across the terminals?

Prob. 42-3. A copper bus-bar distributing power in a substation conducts a current of 400 amperes. The potential difference from one end of the bus to the other is 0.3 volt. What is the resistance of the bus?

Prob. 43-3. In a large motion-picture projector the arc lamp conducts a current of 6.5 amperes with a potential of 85 volts across the arc to make it burn with a steady light. What resistance must be added in series with the lamp in order to use it on a 115-volt circuit?

Prob. 44-3. When a projection arc lamp is burning properly, its resistance is 12.3 ohms and its current 6.5 amperes. How much resistance must be added in series with the lamp for use on 110 volts?

Prob. 45-3. Two resistors connected in parallel have an equivalent resistance of 4 ohms. One of the branches of this parallel combination has a resistance of 16 ohms. What is the size of the other resistor?

Prob. 46-3. If a current of 6 amperes is to be set up in the 16-ohm branch of Prob. 45-3,

(a) What potential will be required across the combination?
(b) What will the current in the other branch be?

Prob. 47-3. The average resistance of the human body is 10,000 ohms. About 0.1 ampere through the body is usually fatal. What is the lowest voltage which would ordinarily kill a person?

Prob. 48-3. A potential of 3.9 volts is required to set up a current of 40 amperes in a length of wire 5 miles long. What is the resistance of the wire per mile?

Prob. 49-3. What potential would be required to set up a current of 60 amperes in an 8-mile length of the wire in Prob. 48-3?

Prob. 50-3. How much resistance must be added in series with a heater operated from 110 volts in order to reduce the heater current from 2.18 amperes to 1.54 amperes?

Prob. 51-3. In Fig. 32, the potential between p and q is 60 volts.
(*a*) What is the voltage between A and D?
(*b*) What is the voltage between A and p?

Prob. 52-3. In Fig. 32, what voltage would have to exist at the terminals of the generator to set up a current of 8 amperes in the resistors?

Prob. 53-3. With reference to Fig. 32, what would be the combined resistance of the three resistors, if they were joined in parallel?

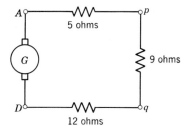

Fig. 32. The generator supplies three resistors in series.

Prob. 54-3. If the parallel combination of Prob. 53-3 were to have a current of 8 amperes entering it, what voltage would be required?

Prob. 55-3. Determine the current in each resistor in Prob. 54-3.

Prob. 56-3. What is the equivalent conductance of the three resistors in Prob. 53-3?

Prob. 57-3. For the three resistors shown in Fig. 32, compute their equivalent conductance when they are connected in series.

4
Magnets and Magnetism

1. **Magnetic Principles.** Many electrical signaling, measuring, and controlling devices depend on magnets of one kind or another for their action. It is essential, therefore, before taking up the study of these devices, to learn some of the facts about magnets and magnetism.

When pieces of iron, nickel, or other iron-like substances are acted upon by certain other iron devices with or without coils carrying electric current, the force experienced in such iron parts is called a *magnetic force*. A piece of iron that can attract other pieces of iron is said to be *magnetized*. If a certain piece of iron can continue to attract other pieces of iron over a long period of time without any outside help, the piece of iron is said to be *permanently magnetized,* or to be a *permanent magnet.* If the piece of iron attracts other iron only as long as the piece of iron has electric current in a coil of wire around itself, the magnet is said to be *an electromagnet.* Both permanent magnets and electromagnets are employed to give useful forces in electrical devices.

There are two contributing causes to these useful magnetic forces. In any magnetic device, whether it has a permanent magnet or an electromagnet in it, the magnetic action is set up by a kind of magnetic urging, called *magnetomotive force.* This magnetomotive force penetrates or permeates the material on which it acts, and this magnetic urging creates lines of action, or *flux lines,* along which the mechanical force of magnetism actually exists. If a strong mechanical force due to magnetism occurs, we usually say that this large force resulted because the magnetomotive force caused a large number of flux lines. If the resulting mechanical forces due to magnetism are weak, when set up by the same magnetomotive force, we say that the magnetomotive force caused only a small number of flux lines. In other words, the magnetomotive force describes the strength of the *attempt* to bring about magnetization, and the number of flux lines describes the strength of the actual result. Since more mechanical force is generally experienced when a magnetomotive force permeates iron instead of air, we usually say that iron has greater *permeability* than air. Thus, permeability gives a description of the strength of magnetic flux that can be set up in a material by a certain magnetomotive force.

When a bar of hard steel is placed in a coil of wire and a current is set up in the wire by a battery or other source, the steel becomes magnetized. After the steel is removed from the magnetomotive force of the coil, the bar will be found to retain a large portion of its magnetism. In other words, the bar becomes a permanent magnet. Needles of magnetic compasses, the small magnets used in magnetic memo pads, and the magnets used to hold refrigerator and cabinet doors closed are examples of permanent magnets.

For either an electromagnet or a permanent magnet, the end from which flux lines leave is called the *north pole* of the magnet. The end where the flux lines enter is called the *south pole*. There is a polarization rule: *Like poles repel each other, and unlike poles exert forces of attraction on each other.* Notice the similarity with the polarity rule for electric charges in Section 2-7.

If a bar of permanent-magnet material is bent around into the shape of a horseshoe or U and then magnetized, the permanent magnet that results is called a horseshoe magnet. If two horseshoe magnets are pushed toward each other so that like poles are facing, the magnets repel each other in accordance with the polarization rule. Figure 33

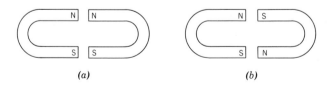

(a)   (b)

Fig. 33. (a) Two horseshoe magnets with like poles close together repel each other with increasing force as they are brought closer together. (b) Two horseshoe magnets with unlike poles near together attract each other with increasing force as they come closer together.

shows both possible orientations of both sets of poles for two horseshoe magnets.

2. Flux Lines. The lines of action along which magnetic forces act are called lines of magnetic flux, or flux lines, as noted in the previous section. For many magnetic devices, it is helpful to picture the behavior in terms of these flux lines. Therefore, it will be useful to look at some of the ways in which flux lines themselves behave.

From many observations on magnets it is possible to state some rules for the behavior of flux lines:

Rule A—When traced through all intervening materials, magnetic flux lines form closed loops.

Rule B—A flux line never crosses itself or another flux line.

Rule C—When flux lines or magnetic materials in their path are in motion, the flux lines stretch or contract in an elastic manner after the fashion of rubber bands.

Rule D—Flux lines that pass near each other headed in the same direction exert forces of repulsion on each other.

This action of flux lines upon each other just stated as Rule *D* is illustrated in Fig. 34. The lines may move apart because of these repelling forces, stretching according to Rule *C*, but the lines must keep an arrangement that will leave them in loops according to Rule *A*, without crossing anywhere in accordance with Rule *B*. With two bundles of flux lines, if each bundle is headed in opposite directions, the two patterns are attracted and merge into some new pattern that satisfies Rules *A*, *B*, and *C*.

When a sketch is drawn of the flux lines for a magnetic device, one other property should be noted. In those parts of the magnetic device that have large numbers of flux lines crowded closely together, the magnetic forces are strongest. Weak magnetic forces occur where the magnetic flux lines are more widely separated. In the usual magnetic devices the most dense tufting of flux lines occurs in narrow iron parts, with the lines spreading out into loose patterns through long paths in

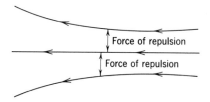

Fig. 34. Repulsion of like-directed flux lines.

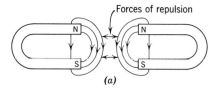

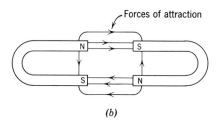

Fig. 35. Flux patterns for two horseshoe magnets: (*a*) like poles repelling; (*b*) unlike poles attracting.

air. The difference in permeability between air and iron paths for magnetic flux results in another rule:

Rule E—For sketches of magnetic devices involving air and iron in the magnetic path, the flux lines enter and leave the iron at right angles to the surface of the iron.

For the horseshoe magnets of Fig. 33, some lines of flux have been sketched in Fig. 35. Note how the elastic behavior of the lines of flux can be thought of as drawing the two magnets together when unlike poles are near each other. Also note how the flux lines repel each other, repelling the magnets which are the sources of the lines, when like poles are opposed to each other.

In Fig. 36, a horseshoe magnet acts on a bar of unmagnetized soft iron. There is a force of attraction drawing the iron to the magnet. Permanent magnets attract unmagnetized magnetic materials by causing

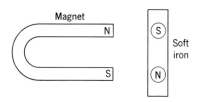

Fig. 36. The soft-iron bar is attracted to the magnet.

them to be temporarily magnetized while in the presence of the force lines of the permanently magnetized material. The letters N and S, shown inside circles on the bar of soft iron in Fig. 36, designate the poles that are temporarily formed in the soft iron because of the nearness of the permanent magnet. The force of attraction may then be deduced from the polarization rule because unlike magnetic poles are facing each other.

Prob. 1-4. Sketch some of the magnetic flux lines around a straight bar magnet. Indicate the poles by the letters N and S.

Prob. 2-4. Sketch some of the flux lines for the horseshoe magnet and soft iron bar of Fig. 36. Explain the attraction of the magnet for the bar in terms of the flux lines, rather than in terms of the polarization rule.

3. The Compass Needle. A small magnet mounted on a pivot turns its north pole toward the Arctic regions of the earth. It may be stated that the earth has magnetic lines leaving its surface in the Antarctic regions, passing more or less in a sheath all around the globe, and returning into the earth in the Arctic regions. A sketch of a few of these lines is shown in Fig. 37 as the lines might appear if a compass or other magnetic disturbance were not present. A compass needle is also shown in a position not lined up with the earth's flux lines. Around the compass are shown a few of the flux lines due to the permanently

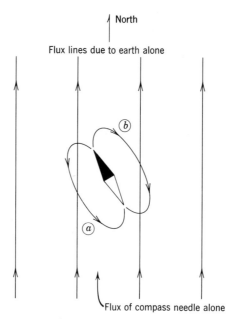

Fig. 37. These two flux patterns do not occur together.

4: Magnets and Magnetism 43

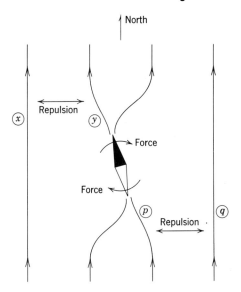

Fig. 38. Flux pattern near deflected compass needle.

magnetized needle itself, in the positions they might have if the needle could be removed from the flux lines of the earth. In the space around *a* and *b* there are flux lines that are headed in opposition; there are crossings of flux lines. This pattern will not exist, rather, changes will take place so that the magnet's lines and the earth's lines will be merged into a pattern such as the one sketched in Fig. 38.

The earth's flux lines, shown passing through the steel in Fig. 38, are merged with the lines of the compass needle. Grouped at *y* is a pair of flux lines, distorted toward *x* by the location of the compass needle. Since the line shown at *x* and the two shown at *y* are headed the same way, there is a force of repulsion in accordance with Rule *D*. A similar condition may be noted at *p* and *q*. By this means a turning effort is exerted on the compass needle, swinging it around until the repulsive forces of flux lines on each side are balanced, as shown in Fig. 39.

The compass needle may be employed to detect the presence of magnetic forces. The regions in or near a magnetic device where these forces occur are said to be regions in which a *magnetic field of force* exists. This space where the flux lines exist is often simply called a *magnetic field*. If many flux lines are closely crowded together, the forces will be strong. Both the presence and the form of magnetic fields that are stronger than the force field due to the earth's magnetic flux may be observed with the aid of a compass needle. The forces of attraction or repulsion due to the strong field will move the compass needle from its

44 Essentials of Electricity

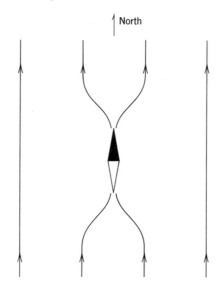

Fig. 39. Flux pattern near compass needle when pointing north.

northerly heading, aligning the needle with the flux lines of the strong field, with the north pole of the compass needle pointing in the direction of the flux lines of the strong field.

Prob. 3-4. Sketch the earth as a sphere, and label two diametrically opposite points as the Arctic and Antarctic regions. Sketch a compass needle pointing north at some point in the northern hemisphere. Label the earth and the compass needle with the appropriate north- and south-pole designations in terms of the definitions given in Section 4-1. With the aid of the polarization rule, explain the force that caused the compass needle to swing around until it pointed north.

Prob. 4-4. A large steel plate stands vertically so that a horizontal line along it runs north and south. A man holding a pocket compass stands on the east side of the plate and brings the compass up to the plate. In what direction does the compass needle point? Explain your answer. In what direction does the compass needle point if the plate is aluminum?

4. Electromagnets—The Right-Hand Rule. The flux lines of a magnetic field are set up by magnetomotive force. In common usage, the measure of magnetomotive force is expressed in terms of a quantity called the *ampere-turn*. We do not consider this unit of measure for permanent magnets because the magnetomotive force in them is the result of complicated arrangements of very many small forces within the actual molecules of the permanently magnetized material. However, we do look at the ampere-turn as a measure of magnetomotive force for magnetic devices that have coils of wire, i.e., electromagnets.

4: Magnets and Magnetism

If several turns of wire are wound around a hollow cardboard tube, and the terminals of the coil of wire thus formed are connected to a battery as shown in Fig. 40, there will be an electric current in the wires around the tube in the definite direction indicated by the arrowheads on the wires. As long as there is current in the wires, we find that all around the coil, both inside and outside, there exists the magnetic effect we have described as lines of action along which mechanical forces exist due to magnetism. These flux lines may be detected by means of a magnetic compass.

Suppose we place a small pocket compass at point A, near the left end of the tube a little above the center line, and then move the compass in the direction in which the north pole of its needle points, continuously changing the direction of motion as the needle swings around. We find that this procedure takes the compass around a path indicated by the dotted line, returning the compass inside the tube to the starting point. If we had started at a point B just below the center line, the path would have extended around the lower part of the tube rather than the upper part. These lines along which the exploring magnet points are the lines of magnetic flux set up by the current in the coil, and the direction of the lines is indicated by the north pole of the magnetic needle. Since the current is directed one way and the magnetic flux lines another, with both directions in space rather than in a plane, it is good to have a simple rule for remembering these directions:

Right-Hand Rule—The direction of magnetic flux lines is at right angles to the direction of the current that caused the flux lines, in the same kind of right-angled relation that the thumb and the fingers of the right hand bear to each other.

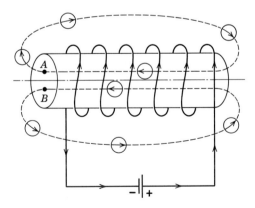

Fig. 40. Dashed lines show flux paths through a coil of current-carrying wire.

46 Essentials of Electricity

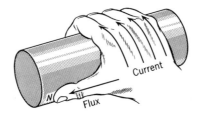

Fig. 41. Right-hand rule applied to a coil of wire.

Applying this rule to the flux lines through the cardboard tube will show how the rule is used. If we grasp the coil with our right hand in such a way that our fingers point along the turns of wire in the same direction as the current, then our thumb points in the direction of the magnetic flux lines through the middle of the coil. Figure 41 shows how the right hand is held to illustrate the right-angled relation between magnetic flux and current.

Recall that the measure of magnetomotive force is in units called ampere-turns. This name shows that magnetomotive force is dependent upon two quantities, the current in the wires and the number of turns in the coil carrying the current. Usually in American practice, magnetomotive force is found by multiplying the size of the current in amperes directly by the number of turns in the coil. Therefore, it is possible to produce the same magnetomotive force by a few turns carrying a large current as would be produced by a coil of many turns carrying only a small current.

Example 1. For a coil produced by winding 200 turns of wire on a cardboard form, what is the magnetomotive force when the current in the coil is 0.5 ampere?

$$\text{ampere-turns} = \text{amperes} \times \text{turns}$$
$$= 0.5 \times 200 = 100 \text{ ampere-turns}$$

Example 2. What magnetomotive force is produced when a coil of 1000 turns of fine magnet wire, wound on a soft-iron core, carries a current of 0.1 ampere?

$$0.1 \times 1000 = 100 \text{ ampere-turns}$$

These two examples show that a magnetomotive force of 100 ampere-turns can be produced, either by 200 turns carrying 0.5 ampere, or by 1000 turns carrying 0.1 ampere. The resulting magnetic effect, or number of flux lines, would be the same in both cases if the dimensions of the coils were the same and if the wires were surrounded by the same substance.

However, in Example 1 the path for flux through the center of the coil is in air, whereas in Example 2 the path is through soft iron. The number of flux lines, or magnetic effect, is much greater for the coil with the iron core. Since the iron has greater permeability than air, the same magnetomotive force sets up more flux in an iron core than when it acts on a core of air. High magnetic permeabilities are obtained with irons and steels, and some alloys with nickel and cobalt, the chief materials for electromagnet cores.

Prob. 5-4. An electric buzzer has two coils wound on soft iron cores as shown in Fig. 42. The piece of iron A is called the armature and is attracted by the electromagnets. Sketch some of the lines of magnetic flux due to the current in the coils.

Prob. 6-4. Which would produce the greater magnetizing effect when wound on the same soft iron core, 100 turns carrying 5 amperes or 1500 turns carrying 0.2 ampere? What is the magnetomotive force in each case?

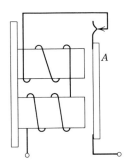

Fig. 42. Electric buzzer.

5. Forces of Magnetism. Because the usefulness of magnetic devices centers around the forces of magnetism, it is important to be able to describe these forces. The polarization rule, stating that unlike poles attract, offers one scheme for describing magnetic forces. The rules about flux lines give a more detailed description. It is possible to visualize the forces as being set up by several different types of interactions of flux lines due to permanent magnets, electromagnets, and even the various parts of the molecules of magnetic materials located near other magnetized materials. In fact, even for devices that employ current-carrying wires placed in magnetic fields it is still possible to visualize the behavior in terms of the rules concerning flux lines. Although this flux-interaction picture can always be applied, it is often helpful to separate magnetic devices into two classes: (1) those in which the forces of attraction or repulsion occur between materials either permanently or temporarily magnetized, and (2) those in which magnetic forces occur when nonmagnetic current-carrying wires are located in regions where flux lines exist, i.e., in a magnetic field.

Forces of the first kind are the forces that are at work between two permanent magnets such as the horseshoe magnets of Fig. 33, or the forces between the horseshoe magnet and the soft-iron bar of Fig. 36. However, it is not necessary that the magnetomotive force be created by a permanent magnet. In doorbells, the magnetomotive force is set up by the coils of an electromagnet. The resulting lines of force lie

48   Essentials of Electricity

Fig. 43. Lifting magnet (55-inch diameter) in action. *Cutler-Hammer Inc.*

along a path through a movable piece of magnetic material, magnetizing it temporarily in much the same fashion as the soft-iron bar of Fig. 36 is temporarily magnetized. The forces between the magnetic poles that temporarily appear on the movable doorbell clapper arm and the poles of the electromagnet are forces of attraction, in accordance with the polarization rule. The clapper is drawn up by these forces, usually working against a spring, to strike the bell. A similar action occurs in the buzzer of Fig. 42, the spring action of the armature itself producing the vibration that generates the buzzing sound. The lifting magnet of Fig. 43 is another example of a device in which the magnetomotive force is produced by an electric current in an electromagnet, rather than by a permanent magnet.

Magnetic forces of the second kind involve the reaction of one set of flux lines due to a current with another set of flux lines due to some other source of magnetism, whether a permanent magnet or an electromagnet. These forces of the second kind involve further applications of the right-hand rule. Recall that the right-hand rule was previously stated as: The direction of magnetic flux lines is at right angles to the direction of the current that caused the flux lines, in the same kind of right-angled relation that the thumb and the fingers of the right hand bear to each other. The hand sketched in Fig. 41 demonstrated one application of the rule. Another application of the rule will show the direction of magnetic flux lines set up by an electric current in a long straight wire.

6. Current-Carrying Wire. The sketch in Fig. 44 shows a long straight wire connected to a battery. Several lines of magnetic flux are shown around various parts of the circuit. Along the straight section of wire, the lines are circles in planes at right angles to the direction of the wire. The direction of magnetic urging, which determines the direction of an arrowhead placed on a flux line, is given by the right-hand rule applied in a new way. As shown in Fig. 45, the thumb of the right hand is placed as though it were pointing in the same direction as the current in the wire. Then the fingers point in the direction to be assigned to the flux lines. Be careful to distinguish the two situations: for a coil,

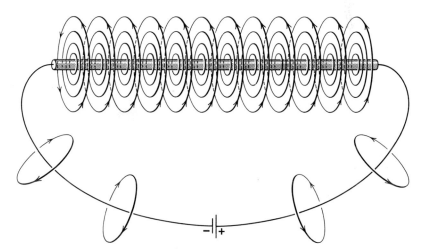

Fig. 44. Magnetic field about a straight wire carrying current.

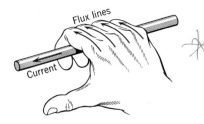

Fig. 45. Right-hand rule applied to a straight wire carrying current.

the fingers represent the current while the thumb represents the flux; for a straight wire, the thumb represents the current while the fingers represent the flux.

The circles of magnetic flux around a straight wire are shown another way in Fig. 46. A battery sends a current through the wire piercing a sheet of cardboard arranged parallel to the ground with the earth's north pole in the direction marked. As employed to detect the flux inside a coil of wire, a small pocket compass can determine the shape of the flux lines of the magnetic field around the wire. If the current is large enough that the magnetic field of the current is much stronger than the field of the earth, the compass needle will lie along the flux lines of the field due to the current. If the compass is placed on the cardboard and moved continuously in the direction its needle points, the compass will trace out a circle around the wire. If the compass is started at another location, as shown in Fig. 46, another circle will be traced out around the first circle. In this way the arrangement of the flux lines around the wire due to the current in this straight wire can be seen to be a concentric pattern of circles. As noted in

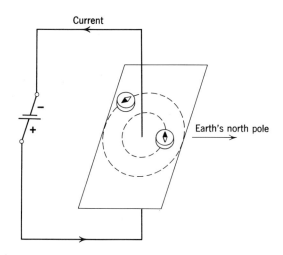

Fig. 46. Magnetic field about a wire shown by compass needles.

Section 4-2, a sketch of magnetic lines shows the largest number of lines where the forces are strongest. Since the forces of the magnetic field around the wire are strongest near the wire, a sketch of the flux lines should show more circles near the wire and fewer farther away.

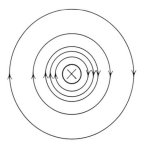

Fig. 47. View of magnetic field about a wire with current directed into the page.

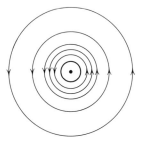

Fig. 48. The reverse of Fig. 47, i.e., current directed out of the page.

Such sketches of the flux lines around a wire are shown in Figs. 47 and 48. If the flux lines were viewed from the under side of the cardboard in Fig. 46, the pattern of Fig. 47 would be seen. The cross in the center of the wire is often employed to mean that the current in the wire pierces the sketch going away from the reader. Similarly, the dot in the center of Fig. 48 means that the current in the wire pierces the sketch to come upward toward the reader. If an arrow is visualized *inside* the wire pointing in the direction of the current, the dot represents the appearance of the head of the arrow; the cross stands for the appearance of the feathered tail of the arrow going away. In these terms, the sketch of Fig. 48 would be seen by looking down on the top of the cardboard in Fig. 46.

When a current-carrying wire lies in a magnetic field, the wire and field exert a force on each other. Figure 49 shows a wire without current lying between the pole pieces of a horseshoe magnet. With no current in the wire, most of the flux lines will be straight lines, as shown, between

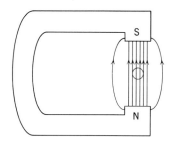

Fig. 49. Flux pattern of horseshoe magnet when there is no current in the enclosed wire.

52 Essentials of Electricity

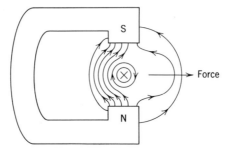

Fig. 50. Distortion of flux pattern with a current-carrying wire threaded into an otherwise straight magnetic field.

the pole pieces. With current in the wire, in a direction going away from the reader, as shown by the cross in the wire of Fig. 50, the flux lines due to the current would be circles if they could exist alone. However, these flux lines would have crossings with the flux lines of the magnet, and there would be other patterns that could not exist, in accordance with Rule D of Section 4-2. Therefore, the pattern of Fig. 50 shows the arrangement of flux lines that would be disclosed by test when both the flux of the current and the flux of the magnet are acting simultaneously. It can be seen that the flux lines are crowded closely together on the left of the pattern and spread loosely on the other side. This crowding will lead to a force of repulsion to rearrange the pattern, thereby exerting a force on the side of the wire to move it to the right. Because this type of magnetic interaction is the basis for developing useful forces in electric motors, the action of a magnetic field on a current-carrying wire is often called *motor action*.

Examination of Fig. 50 shows that the flux lines of the magnet are directed one way, the current in the wire is at right angles to the direction of the flux lines of the magnet, and the force of the motor action is at right angles to both of these other directions. Here is an outline of a simple method to determine the direction of the force correctly: (1) sketch the lines of force that would exist with only the magnet present; (2) add to this sketch the lines of force due to the current in the wire alone, as determined by application of the right-hand rule; (3) the force of motor action will be directed *away* from regions where the two fluxes head in the same direction *toward* regions where the two fluxes merge into new patterns because they are headed in opposite directions.

When a current-carrying wire is subjected to a force of motor action,

4: Magnets and Magnetism 53

the force with which the wire tends to move depends upon the strength of the magnet and the size of the current in the wire. If the current in the wire is doubled, the force tending to move the wire in the magnetic field is doubled. If the current in the wire remains the same, while the strength of the magnetic field per square inch of area is doubled, the force is again twice as large as its original size. Usually the strengths of magnetic fields are changed through adjusting the size of the current in the coils of electromagnets that create the field.

Prob. 7-4. Near the north pole of the bar magnet shown in Fig. 51 is a wire carrying a current of 10 amperes in the direction indicated. Determine the direction of the force exerted on the wire.

Fig. 51. Motor action causes a force on the wire.

Prob. 8-4. Describe how the force would differ in Prob. 7-4 if the current were 2 amperes in the direction opposite to the one indicated in Fig. 51.

Prob. 9-4. An electromagnet of the type employed in buzzers and doorbells is shown in Fig. 52, with a current of 0.2 ampere in its coils in the direction indicated. A soft-iron bar is mounted nearby so that it cannot move. Determine the correct markings of north and south for the two pole pieces of the electromagnet. Mark the names of the temporarily magnetized poles that appear in the soft-iron bar.

Prob. 10-4. There is a total of 1200 turns for the two coils shown in Fig. 52. Determine the size of the total magnetomotive force active in setting up flux due to the electromagnet alone.

Prob. 11-4. The two current-carrying wires marked A and B in Fig. 52 each have a current of 3 amperes in the directions shown by the dots. Determine the direction of the forces of motor action that are exerted on each wire.

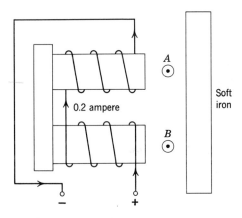

Fig. 52. The soft-iron bar does not move.

54 Essentials of Electricity

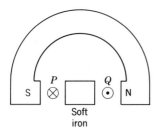

Fig. 53. A pair of current-carrying wires in a magnetic field.

Prob. 12-4. A horseshoe magnet has a block of soft iron fastened in the space between its poles, as shown in Fig. 53. Two wires shown as P and Q, each carrying 0.1 ampere in the sense marked, are directed at right angles to the flux lines. Determine the direction of the forces on each of the wires.

SUMMARY

A piece of iron that can attract other pieces of iron is said to be MAGNETIZED. PERMANENT MAGNETS remain magnetized for long periods of time, whereas ELECTROMAGNETS are temporarily magnetized by electric currents in their coils of wire.

The action behind magnetism is called MAGNETOMOTIVE FORCE; for electromagnets this quantity is measured in AMPERE-TURNS.

The action of magnetomotive force sets up a number of FLUX LINES, or lines of magnetic forces, the number of lines depending upon the PERMEABILITY of the path of the forces.

Flux lines LEAVE the NORTH pole of a magnet, and ENTER the SOUTH pole.

LIKE poles REPEL, and UNLIKE poles ATTRACT.

Magnetic flux lines always form LOOPS which stretch and contract in an ELASTIC MANNER after the fashion of rubber bands.

In reasonably strong magnetic fields, a POCKET COMPASS will lie pointing along the flux lines.

Electric-current direction and flux-line direction are related by the RIGHT-HAND RULE:

The direction of magnetic flux lines is at right angles to the direction of the current that caused the flux lines, in the same kind of right-angled relation that the thumb and the fingers of the right hand bear to each other.

Magnetic forces may be divided into TWO CLASSES: (1) those acting between MAGNETIC MATERIALS, and (2) those acting between a magnetic field and a CURRENT-CARRYING WIRE.

Permanent magnets attract unmagnetized magnetic materials by TEMPORARILY MAGNETIZING them.

The force of MOTOR ACTION is at RIGHT ANGLES to BOTH the directions of the current and the magnetic field that jointly produce the force.

The SIZE of the force of MOTOR ACTION depends on the size of the CURRENT and the strength of the MAGNETIC FIELD.

5

Meters and Instruments

1. Definitions. In Chapter 1 the basic units for describing electricity in motion were defined. Current is measured in amperes, and potential is measured in volts. However, the nature of electricity is such that we cannot judge the size of its various measures without the use of relatively complicated devices that indicate the desired quantity. Such devices are called electric instruments or electric meters, or often simply either instruments or meters. As defined in the dictionary, an *instrument* is a device that provides a measurement of a quantity immediately, at the moment of observation. On the other hand, the dictionary defines a *meter* as a device that totalizes or records the sum of all values of a quantity measured by the device over a period of time. In this sense, a meter is a device that registers the product of the quantity to which it responds and the time of the response.

Common usage has mixed these meanings. The device that measures the resistance of a circuit measures ohms but certainly does not totalize ohms over a period of time. Nevertheless, electrical workers know such a device by the name of ohmmeter; the name ohm-instrument would not be clear. Furthermore, a sense of accuracy or precision has become associated with the word instrument. Although there are many mass-produced devices called voltmeters and although many of these voltmeters may possess high accuracy, the voltage-measuring devices employed in testing laboratories are usually called instruments, in keeping with the dictionary definition. Another aspect of this mixed usage is that many manufacturers of voltmeters, ammeters, and ohmmeters have the word "instrument" in the name of their company or division.

Although the devices discussed in this chapter belong under the heading of instruments in keeping with the meanings of the dictionary, the common-usage designation of meter is also employed. A device that totalizes energy is referred to in Chapter 7.

2. Convenient Units. In many designs of meters and instruments, the actual current that brings about the operation of the device is a small current. Furthermore, the voltage needed for operation is often also small. Consequently, it is convenient to have smaller measures of current and potential than the usual units of amperes and volts. One of the most common subdivisions of these basic units is a unit just $\frac{1}{1000}$ of the original size. In keeping with the international usage of the metric system, a prefix of "milli," implying $\frac{1}{1000}$, is placed before the name of the original unit. Therefore, 1 *milliampere* is a unit of current equal to $\frac{1}{1000}$ of an ampere, or 1 milliampere equals 0.001 ampere. Similarly 1 *millivolt* is a unit of potential equal to $\frac{1}{1000}$ of a volt, or 1 millivolt equals 0.001 volt. Note that, because the factor of definition is $\frac{1}{1000}$ in each case, dividing millivolts of potential by milliamperes of current does not change the ratio. As a result, dividing the potential in millivolts by the current in milliamperes still gives the resistance of a portion of a circuit in ohms.

Example 1. The coil of a very small electromagnet requires 0.5 volt across it to establish a current of 0.025 ampere through it. What is the current in milliamperes? What is the potential in millivolts? What is the resistance of the coil in ohms?

The potential is $\frac{500}{1000}$ volt, or 500 millivolts.
The current is $\frac{25}{1000}$ ampere, or 25 milliamperes.

$$\text{resistance} = \frac{\text{voltage}}{\text{current}} = \frac{0.5 \text{ volt}}{0.025 \text{ ampere}} = 20 \text{ ohms}$$

$$= \frac{500 \text{ millivolts}}{25 \text{ milliamperes}} = 20 \text{ ohms}$$

3. The Milliammeter. One type of current-measuring instrument that finds common use is known as a D'Arsonval movement, or more descriptively as a permanent-magnet moving-coil movement. Since the device usually is made to be operative with currents that are small fractions of 1 ampere in the moving coil, this instrument is known as a *milliammeter*. An outline of the essential elements of a milliammeter is shown in Fig. 54, and a sketch of the moving coil is given in Fig. 55. The permanent magnet arranged in the form of a horseshoe provides a

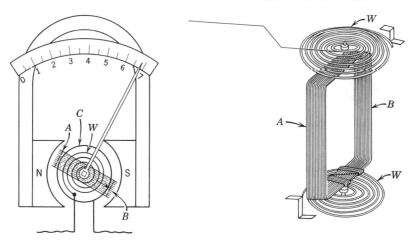

Fig. 54. Diagram of a milliammeter.

Fig. 55. Moving coil of a milliammeter.

constant magnetic field through the pole pieces marked N and S, and through a central core C of soft iron. Arranged on pivots to turn around this core is the bobbin carrying the coil of wire shown in Fig. 55. The sides of this coil of wire, marked A and B, are the current-carrying wires that develop a force through the agency known as motor action. The force developed by motor action causes the bobbin to turn on its pivots against a restraining force developed by the two coil springs W. Electrical connections are made to the two mounting tabs on the ends of springs W so that contact with the circuit is not broken when the coil moves. The fact that the current that operates the instrument passes through these watch springs is one reason that the size of the current is in the milliampere range. Figure 56 shows a commercial instrument movement of the horseshoe-magnet kind.

Also shown in Fig. 54 is a pointer attached to the moving coil, with the pointer lying over a numbered scale. The force on the bobbin due to the motor action of the current to be measured by the instrument causes the bobbin to turn until watch springs W develop an equal but oppositely directed restraining force on the bobbin. When the bobbin has reached a position where these two forces are equal, the pointer comes to rest at some location on the scale. With proper shapes for the magnetic pole pieces and the central iron core, and with springs that develop a uniformly increasing restraining force as they are wound up, distances along the scale will be proportional to the current in the moving coil.

58   Essentials of Electricity

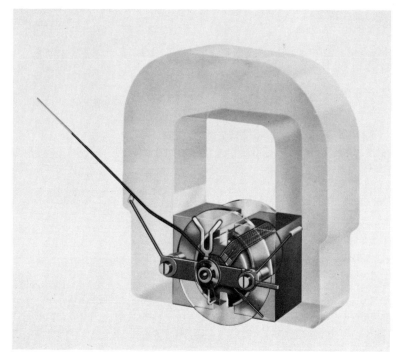

Fig. 56. Permanent-magnet moving-coil mechanism.   *Weston Instruments Div., Daystrom, Inc.*

Because the permanent magnet of a milliammeter sets up magnetic flux in a definite direction, current must enter the instrument in a definite direction if the pointer is to move *upscale* against the spring force. On many instruments this information is indicated with a plus (+) sign on the terminal at which current should enter.

When large numbers of this type of instrument are mass produced, proportionally spaced numbers are printed in advance on an interchangeable scale. For more accurate laboratory instruments, a blank scale is marked by hand, with several points actually checked against the reading of a standard instrument. The instrument to be *calibrated*, the name by which this procedure is called, is fitted with a blank scale and inserted into a series circuit with the standard instrument S as shown in Fig. 57. The device at R is an adjustable resistor, sometimes called a rheostat. Rheostat R is adjusted until the current in the circuit causes the pointer of the standard instrument to lie exactly at one of its principal markings, perhaps at 40 milliamperes on a 100-milliampere range. A point is marked where the pointer lies on the blank

5: Meters and Instruments   59

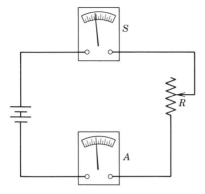

Fig. 57. Standard instrument S is arranged for calibrating working instrument A.

scale of the instrument A being calibrated. This procedure is repeated for other principal, or cardinal, points on the standard instrument that lie within the range of the instrument undergoing calibration. Various methods are then employed to subdivide the whole-number divisions into smaller fractions.

Instruments of this type are usually built to be so sensitive that a very small current in the coil will give full-scale deflection. For instance, the current required for full-scale deflection might be 25 milliamperes. In this case the scale would be marked from 0 to 25 and would indicate milliamperes.

Even smaller full-scale currents are possible with the permanent-magnet moving-coil movement. Many mass-produced milliammeter movements are available with a full-scale current of only 1 milliampere, and still others are readily available with a full-scale current of 0.05 milliampere. These low-current instruments, however, usually do not possess the accuracy or precision of the less-sensitive hand-calibrated laboratory instruments.

Prob. 1-5. A certain milliammeter has an internal resistance between its terminals of 9 ohms. For full-scale deflection it requires a current of 25 milliamperes in its coil. What is the potential across the terminals of the milliammeter when the pointer is at full scale?

Prob. 2-5. What would the current be in the milliammeter of Prob. 1-5 when the pointer is at only 20% of full scale?

Prob. 3-5. If the current in the milliammeter of Prob. 1-5 is 20 milliamperes, what is the voltage across the terminals of the meter?

Prob. 4-5. For the conditions of Prob. 3-5, at what percentage of full scale does the pointer of the milliammeter lie?

Prob. 5-5. For the milliammeter of Prob. 1-5, what current will exist in its coil when a potential of 50 millivolts is applied across the instrument terminals?

Prob. 6-5. A D'Arsonval instrument requires 100 millivolts for full-scale deflection, with the full-scale indication being 150 milliamperes. What is the resistance of this meter?

Prob. 7-5. What current will exist in the coil of the milliammeter of Prob. 6-5 if a voltage of 30 millivolts is impressed across its terminals?

Prob. 8-5. If the current in the milliammeter of Prob. 6-5 is 50 milliamperes, what is the potential across its terminals? At what fraction of full-scale deflection will the pointer lie?

4. The Ammeter. The milliammeter, described in the previous section, can be used to measure any magnitude of direct current if it is provided with a *shunt*. The shunt is a parallel resistor, usually having very small resistance, which is connected across the terminals of the basic milliammeter. The parallel resistor may be thought of as affording another path for current, diverting or shunting it away from the sensitive milliammeter. Figure 58 shows how a shunt is connected to a milliammeter, with a switching link K indicated as a means of disconnecting the shunt when it is not wanted. It should be clearly understood that the shunt and the instrument movement merely form a simple parallel circuit, to which the rules of Chapter 3 apply just as well as to any other parallel electric connection.

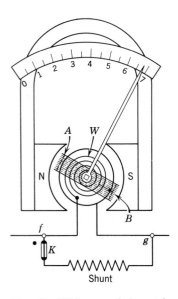

Fig. 58. With part of the total current in the shunt, the range of the milliammeter is increased because of the shunt.

The resistance of the shunt is chosen of a size such that it takes a certain proportion of the total current to be measured, thus causing full-scale deflection of the milliammeter with its own original full-scale current while the total current of the entire circuit is usually considerably larger than that of the milliammeter alone. For example, suppose that the meter movement of Prob. 1-5 is provided with a 9-ohm shunt. Since the resistance of the milliammeter is also 9 ohms, the two 9-ohm values of resistance will divide the total current of the circuit equally at point f, with the circuit marked as in Fig. 58. With half of the total current in the shunt and the other half in the milliammeter, the pointer of the meter will be at full scale when the total external current, from f to g in the diagram, is 50 milliamperes. Because Ohm's law is a proportional law, the actual external current is always twice the current in the coil of the milliammeter with this shunt.

As a second example, assume that the shunt has a resistance of 1 ohm, for the 9-ohm milliammeter. Full-scale deflection still occurs when the current in the meter bobbin is 0.025 ampere. By Ohm's law the voltage drop across the coils is $25 \times 9 = 225$ millivolts, or 0.225 volt. This potential must be the same as the voltage across the shunt. Therefore, the current in the shunt must be $0.225/1 = 0.225$ ampere, or 225 milliamperes. The total current through the new instrument comprising the 9-ohm milliammeter and the 1-ohm shunt is the sum of the separate parallel-branch currents or $0.025 + 0.225 = 0.250$ ampere, or 250 milliamperes. When the pointer on the milliammeter lies at its full-scale mark, the total current being measured is 250 milliamperes. The connection of the 1-ohm shunt to the 9-ohm movement has increased the range of current that can be measured with the milliammeter by a factor of 10. If the complete instrument has a switching link K as indicated in Fig. 58, two ranges are possible. With the link open (moved to the left in the diagram), the instrument may measure currents in the 0–25 milliampere range. With the link closed (to the right as shown), the instrument then has a 0–250 milliampere range.

If the shunt has a much smaller resistance than those already considered, say $\frac{1}{111}$ ohm, the full-scale reading corresponds to a total current many times that in the coil of the meter movement. For full-scale deflection of the movement of Prob. 1-5, the potential across the terminals of the movement is $0.025 \times 9 = 225$ millivolts as before. But the current in the new shunt is

$$\frac{0.225 \text{ volt}}{\frac{1}{111} \text{ ohm}} = 24.975 \text{ amperes}$$

The total external current is the sum of the currents through the meter coil and the shunt, or $0.025 + 24.975 = 25$ amperes. Because the total current is in the range of amperes, the complete instrument is called an *ammeter*, a contraction of ampere and meter. The name ampere-meter is not used. Thus the basic milliammeter, which had a range of 0–25 milliamperes, is converted by the shunt into an instrument with a range of 0–25 amperes. The total current for the entire instrument is, at any deflection, 1000 times the current in the coil of the milliammeter at that deflection.

Example 2. What is the range of the ammeter formed from the basic 9-ohm milliammeter with a shunt having a resistance of $\frac{1}{11}$ ohm?

62   Essentials of Electricity

For full-scale deflection the voltage across the ammeter is determined by the full-scale voltage across the milliammeter, as before: $0.025 \times 9 = 0.225$ volt. The current in the shunt is

$$\frac{0.225 \text{ volt}}{\frac{1}{11} \text{ ohm}} = 2.475 \text{ amperes}$$

Since the total current is $0.025 + 2.475 = 2.5$ amperes, the ammeter has a range of 0–2.5 amperes.

Notice that in all but the first of these examples, the shunts have been chosen so that the total current is 10, 100, or 1000 times the current in the coil. When the multiplying factor is a multiple of 10, only one scale need be provided on the basic milliammeter, for the user can determine the correct multiple of 10 and then locate the decimal point accordingly after taking the basic reading. If the ranges of an instrument do not differ by multiples of 10, separate scales are usually provided for the different ranges.

From the examples just considered we may deduce the method of finding the resistance of the shunt required to multiply the readings by any desired factor. To multiply by 10, the resistance of the shunt was $9/(10 - 1)$; to multiply by 100, the shunt had a resistance of $9/(100 - 1)$; for a factor of 1000, the resistance needed was $9/(1000 - 1)$; and in general

$$\textit{resistance of shunt} = \frac{\textit{resistance of coil}}{\textit{factor} - 1}$$

If we wish to measure the current in some piece of apparatus in an electric circuit, we must somehow make a break into that portion of the circuit containing the apparatus. The ammeter must be *inserted* into the circuit so that only the charge flowing in the portion we are interested in passes through the meter. Figure 59 shows a parallel connection of a heater resistor and a lamp connected across line wires that

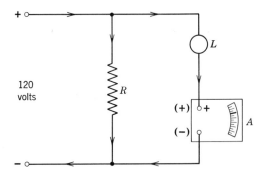

Fig. 59. Ammeter *A* measures the current in lamp *L* only.

provide a potential of 120 volts. An ammeter A is shown inserted only into the path containing the lamp. Therefore, all the current in the lamp is also in the ammeter. It is important that the resistance of the ammeter be small in order not to alter the current in the lamp to a value that would be different from the lamp's current when the ammeter was omitted from the circuit.

Recall from Chapter 3 that the equivalent resistance of two parallel paths is always smaller than the smaller of the two resistances. In Example 2 above, the milliammeter has a resistance (mostly in the coil on its bobbin) of 9 ohms, and the 11-mho shunt has a resistance of only 0.0909 ohm. Note, therefore, that the resistance offered by the coil and shunt of an ammeter is indeed low. Hence caution is necessary in using an ammeter. *Do not connect an ammeter across line wires or any voltage source.*

Note that the terminals of the ammeter shown in Fig. 59 are connected so that the lamp current enters the meter at the plus (+) and leaves at the negative (−) terminal. Commercial ammeters usually have only one terminal marked with a polarity sign, often just the + terminal. This + terminal must be connected toward the positive terminal of the source of potential so that the current in the portion of the circuit where the ammeter is will be directed into the ammeter at its + terminal.

Prob. 9-5. A milliammeter has a full-scale deflection of 100 divisions when the coil current is 10 milliamperes. The coil has a resistance of 15 ohms. (*a*) What size of shunt must be used to give this instrument a range of 0–10 amperes? (*b*) What size of shunt is needed for a range of 0–1 ampere? (*c*) Show a 2-point switch for the purpose of changing from one range to the other.

Prob. 10-5. What would the range of the instrument in Prob. 9-5 be if the shunt had a resistance of 0.0685 ohm?

Prob. 11-5. What would the resistance of a shunt have to be if it were to be used with the ammeter of Prob. 10-5 to establish a new range ten times as great as the ammeter already has in that problem?

Prob. 12-5. A 5-milliampere movement has an internal resistance of 65 ohms. Determine the resistance of a shunt that will convert this movement into an ammeter with a range of 0–50 amperes.

Prob. 13-5. For the milliammeter movement of Prob. 6-5, determine the multiplying factor needed to convert this device into an ammeter with a range of 0–300 amperes. What voltage appears across the shunt when the instrument is measuring 150 amperes?

Prob. 14-5. The milliammeter movement of Prob. 6-5 is to be used as an ammeter with a range of 0–50 amperes. Determine the resistance of the shunt needed.

5. The Voltmeter. In the preceding sections there have been several examples involving a 0–25 milliampere full-scale deflection for a milliammeter having an internal resistance of 9 ohms. We noticed that the

64 Essentials of Electricity

voltage across the movement at full-scale deflection was $0.025 \times 9 = 0.225$ volt. Also, for half-scale deflection there would be $(0.025/2) \times 9 = 0.1125$ volt across the terminals. If then we supply another scale marked *millivolts* in addition to the one marked milliamperes, so that the full-scale point marked 225 millivolts corresponds to the point marked 25 milliamperes, 112.5 millivolts corresponds to 12.5 milliamperes, etc., this instrument can then be thought of as a *millivoltmeter*. In certain circuits it could be used successfully to measure small voltages. Thus we see that the essential difference between a millivoltmeter and a milliammeter is in the scale marking, the basic internal movement having the same general construction for either instrument.

If larger voltages are to be measured, a *series resistor* is connected as shown in Fig. 60. This resistor is often called a *multiplier,* in that its presence with the internal movement greatly increases the range of voltage that can be measured with the given millivoltmeter. Again referring to the basic milliammeter of Prob. 1-5, suppose we wish to convert this movement to a *voltmeter* having a full-scale deflection of 250 volts. The voltage across the movement will still be 225 millivolts at full scale, when the current in the movement is 25 milliamperes. Since the circuit is a series connection between terminals f and h of the voltmeter, the total potential of 250 volts for full-scale deflection must be the sum of the voltages across the meter movement and the series resistor or multiplier. With a total of 250 volts, and with 25 millivolts across the movement, there must be $250 - 0.025 = 249.775$ volts across the series

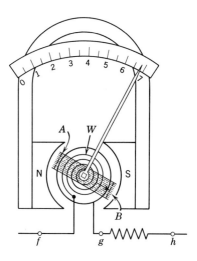

Fig. 60. The milliammeter becomes a voltmeter with addition of a series resistor.

resistor. Hence the size of the resistance must be 249.775/0.025 = 9991 ohms. As a check we may now work backward and find that the current in the instrument, when the terminal voltage is 250 volts, is

$$\frac{\text{volts across instrument}}{\text{total resistance of instrument}} = \frac{250}{9991 + 9} = 0.025 \text{ ampere}$$

This value, of course, is the one needed for full-scale deflection of the movement. Thus we have made the 25-milliampere milliammeter into a 250-volt *voltmeter* by adding a series resistance of 9991 ohms.

From another form of Ohm's law,

$$\frac{\text{volts across instrument}}{\text{full-scale current}} = \text{series resistance} + \text{coil resistance}$$

$$\frac{250 \text{ volts}}{0.025 \text{ ampere}} = 9991 + 9 \text{ ohms}$$

The series resistance to be added to give a full-scale deflection for a certain voltage can be found by dividing this voltage by the full-scale current of the movement and then subtracting from this quotient the coil resistance of the movement; thus

$$\text{series resistance} = \frac{\textit{full-scale volts}}{\textit{full-scale coil current}} - \text{coil resistance}$$

$$= \frac{250}{0.025} - 9 = 9991 \text{ ohms}$$

Example 3. What size should the series resistor have for use with the movement of Prob. 1-5 to give full-scale deflection on 2.5 volts?

$$\text{series resistance} = \frac{2.5}{0.025} - 9 = 100 - 9 = 91 \text{ ohms}$$

Figure 61 shows a small panel meter with a range of 0–5 volts, and Fig. 62 shows a voltmeter with three ranges. Of the four terminals, one is common and each of the others is connected at the appropriate point in a series string of resistors to cause the internal milliammeter to have its full-scale deflection extended by the correct factor to give full-scale readings of 75, 150, and 300 volts.

When we wish to measure the electric potential across a section of an electric circuit, we do not wish to disturb the circuit or the current in it. Instead of *breaking into* the circuit as with an ammeter, we just *attach* the terminals of a voltmeter *across* that part of the circuit, as across the lamp in Fig. 63. In the series circuit shown, a source of about 120 volts creates current in a lamp through two small resistances, perhaps the resistances of wires leading to the lamp. If we want to measure the

66 **Essentials of Electricity**

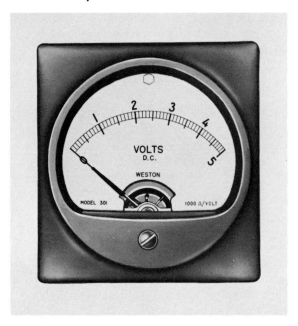

Fig. 61. Small voltmeter for panel mounting. *Weston Instruments Div., Daystrom, Inc.*

potential of the source, we would connect the voltmeter between points *a* and *b*. However, when we want to measure the difference in potential across the lamp, i.e., the voltage that creates the current in the lamp, we attach the voltmeter across the lamp terminals, *c* and *d* without including the two resistances in the path spanned by the voltmeter. Note that the current in the lamp is not found in the voltmeter. A voltmeter must not be placed so that the current of a circuit will enter into it, but it should be attached *across* that part of a circuit on which the potential is desired. As with the ammeter, the polarity markings must be observed so that the forces on the bobbin in the magnetic field of the internal milliammeter move the pointer upscale. The voltmeter should be placed so that its plus (+) terminal is connected to the + side of the lamp, i.e., the side of the lamp on which the current arrow enters.

Some circuits on which voltage measurements are to be made have resistances in the hundreds or thousands of ohms. Because a voltmeter connected across a portion of such a circuit adds a parallel path, it is often important to know the resistance of the voltmeter path, so that an estimate can be made to determine if the voltage on the circuit portion is the same whether or not the voltmeter is connected. On many mass-produced instruments, a figure-of-merit appears so that this estimate can be made easily. In Fig. 61, this value is 1000 ohms per volt, the capital omega (Ω) often being used for ohms. This figure-of-merit means that the voltmeter has a total resistance of its full-scale reading times the figure-of-merit, or for Fig. 61, 5 × 1000 = 5000 ohms. For the voltmeter of Fig. 62, assume that the figure-of-merit (sometimes called the *sensitivity*) is 5000 ohms per volt. On the 300-volt range, this instrument would possess a total resistance of 300 × 5000 = 1,500,000 ohms. On the 75-volt range, the internal resistance would be only 75 × 5000 = 375,000 ohms. Although these values of resistance may

Fig. 62. Three-range testing portable voltmeter. *Weston Instruments Div., Daystrom, Inc.*

68   Essentials of Electricity

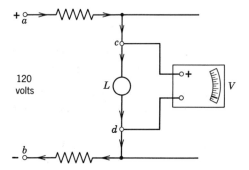

Fig. 63. Voltmeter V measures the voltage across lamp L.

seem high, some circuits require such sizes if connection of the voltmeter is not to disturb the operation of the circuit.

The figure-of-merit for voltmeters may be considered in another light as a specification of sensitivity. If the full-scale voltage of a range is divided by the total internal resistance for that range, the current for full-scale deflection of the internal milliammeter is obtained as shown above. However, since the total resistance may be found by multiplying the full-scale voltage by the figure-of-merit, we could work backward to find the figure-of-merit by dividing the total resistance by the full-scale voltage. For the 300-volt range of the meter of Fig. 62, this calculation would be:

$$\frac{\text{total resistance}}{\text{full-scale volts}} = \frac{1{,}500{,}000}{300} = 5000 \text{ ohms per volt}$$

But this ratio is just the reciprocal of the ratio for the internal current:

$$\frac{\text{full-scale volts}}{\text{total resistance}} = \frac{300}{1{,}500{,}000} = 0.200 \text{ milliampere}$$

Therefore, the full-scale current of the internal milliammeter movement of a voltmeter is the reciprocal of the figure-of-merit in ohms per volt. For the meter of Fig. 62, with the figure-of-merit of 5000 ohms per volt,

$$\frac{1}{5000 \text{ ohms per volt}} = 0.0002 \text{ ampere} = 0.2 \text{ milliampere}$$

Prob. 15-5. What series resistance must be added to the milliammeter of Prob. 9-5 to make it indicate 100 divisions when the external terminals are placed across a 100-volt line?

Prob. 16-5. Specify the range of the voltmeter comprising the milliammeter of Prob. 9-5 and a series resistance of 45.5 ohms.

Prob. 17-5. Show a diagram of connections for series resistors arranged to give a milliammeter two different voltage ranges. What would be the sizes of these series resistors if the movement of Prob. 9-5 is to be equipped for 15-volt and 75-volt scales?

Prob. 18-5. What is the full-scale current required by the voltmeter of Fig. 61?

Prob. 19-5. What series resistor would be required to convert the voltmeter of Fig. 61 into a voltmeter with a range of 0–500 volts?

Prob. 20-5. If the voltmeter shown in Fig. 61 has an internal resistance in its milliammeter of 800 ohms, what is the size of its series resistor to give the 5-volt range shown?

6. Recording Meters. It is often important to keep a continuous record of the voltage or current in certain circuits. Furthermore, if nonelectrical quantities such as aircraft altitude or thickness of newsprint are translated into appropriate analogous electric measures such as voltage or current, continuous records of these other quantities may be obtained in terms of a record of the electric values. A permanent-magnet moving-coil device is the heart of *recording instruments,* or *recorders.* Figure 64(*a*) shows the D'Arsonval movement of a commercial recording meter. The moving coil actuates a pen, shown in Fig. 64(*b*), that is filled from a nonspilling inkwell capable of providing ink for at least

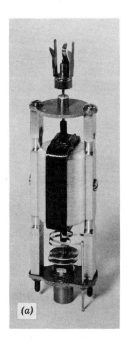
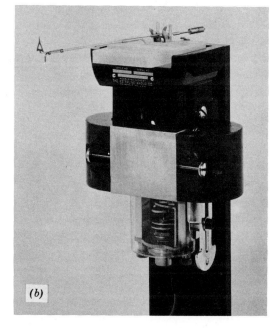

Fig. 64. Assembly for a recording meter: (*a*) moving-coil element; (*b*) moving coil mounted with magnet and pen. *Esterline Angus Instrument Company, Inc.*

70   Essentials of Electricity

Fig. 65. Expanded-scale recording voltmeter. *Esterline Angus Instrument Company, Inc.*

2 months. The chart on which the pen writes, shown in Fig. 65 in an overall view of a recorder, is moved by a motor that maintains the chart feed at a constant rate. Spring motors are often used for the chart drive, either hand-wound or else rewound by an electric motor. A-c motors, called *synchronous* because they are capable of maintaining a fixed number of revolutions over a fixed time interval, can also be fitted to give constant chart speed. With the aid of shunts or series resistances, it is possible to extend the range of the basic milliammeter, just as was shown in Sections 5-4 and 5-5. If the recorder has been fitted as a voltmeter for a range of 0–150 volts, chart paper must be

procured with appropriate markings up to 150 volts. The recorder shown in Fig. 65 has an expanded range that permits covering most of the chart paper with the range from 100 to 140 volts, while the overall range of the recorder is 0–150 volts.

7. Measuring Resistance. Because of the simple proportionality between voltage and current that is expressed by resistance, knowledge of the resistance of a portion of an electric circuit is often the first result desired from electrical testing. The most direct way to find the resistance of a section of a circuit is to measure the current which a known voltage can create in that section and divide the voltage by the current. Connecting a voltmeter and ammeter to a working circuit for this purpose is called the *voltmeter-ammeter method* for measuring resistance.

For example, to find the resistance of lamp L in Fig. 66, an ammeter is inserted into the circuit with the lamp and a voltmeter is attached across the terminals of the lamp. The voltmeter reading is divided by the ammeter reading, the quotient being the desired resistance of the lamp. Note that lamps, as well as other electric devices, may have cold values of resistance differing widely from their hot values. When a lamp is actually operated in a circuit, other controls can set its operation to bring it to any desired temperature. Therefore, the voltmeter-ammeter method is often the simplest method of obtaining the resistance of any electrical device when it is in use in a circuit.

For the voltmeter-ammeter arrangement of Fig. 66, an estimate of current in the voltmeter path may have to be made, as suggested in Section 5-5. If the current in the voltmeter is estimated as 25 milliamperes and the current in the ammeter is 5 amperes, simply dividing the voltmeter reading by the ammeter reading will give an accurate value for the unknown resistance of the lamp. However, if the current to be measured is also in the milliampere range, the observed ammeter reading should be modified. The voltmeter will read the correct volt-

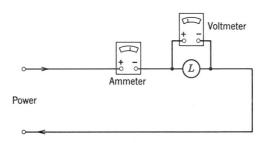

Fig. 66. Connections for the voltmeter-ammeter method for measuring resistance.

age, but the current in the lamp will be the ammeter reading minus the estimated current in the voltmeter. If, for example, the current in the ammeter is only 140 milliamperes and the voltmeter's current is estimated to be 25 milliamperes, the voltmeter reading should be divided by $140 - 25 = 115$ milliamperes for a more correct value of the unknown resistance. The correction can be avoided if a voltmeter with a higher figure-of-merit in ohms per volt is substituted for one that requires current in the voltmeter path comparable to the current being measured in the circuit of Fig. 66.

Sometimes it is helpful to determine the resistance of a portion of a circuit without applying normal voltage. Especially if many measurements in many parts of a circuit are required, the self-contained testing device known as an *ohmmeter* affords great convenience. Figure 67 is a circuit diagram of a basic form of ohmmeter.

Ohmmeters built around this circuit arrangement are known as series ohmmeters because the connection is essentially of the series type. The internal milliammeter M and its adjustable parallel resistor $R1$ are considered as a single element. This metering element is then in series with resistor $R2$, the 1.5-volt battery, and unknown resistance R. The unknown resistance represents the portion of electric circuit to be measured, external to the ohmmeter, and attached to it at its terminals x and y. The scale of such an ohmmeter runs opposite to that of the usual milliammeter, having 0 ohms at the right-hand end and infinitely large ohms (or ∞) at the left. In use, test leads attached to terminals x and y are shorted together and resistor $R1$ is adjusted until the pointer of the ohmmeter is at 0 ohms. This adjustment determines the largest current possible in the leads shorted between points x and y and

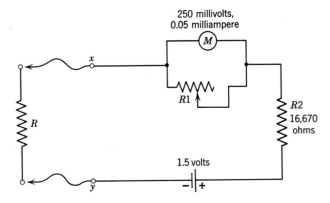

Fig. 67. Simplified circuit diagram of an ohmmeter.

5: Meters and Instruments 73

Fig. 68. Pocket-size multimeter. *The Triplett Electrical Instrument Co.*

brings about full-scale current in the internal milliammeter, which has the 0 marking at its full-scale point. If the test leads are now separated so that they do not touch anything, the series circuit is opened. The meter current becomes zero and the pointer lies at the left-hand end of the scale at ∞, indicating a very high resistance or essentially open circuit. This scale marking appears clearly in Fig. 68, showing a commercial instrument that comprises a voltmeter, ohmmeter, and milliammeter in a single, compact unit, from which the circuit of Fig. 67 has been drawn. Such an instrument is sometimes called a *VOM,* with the initial letters of the key words as an abbreviation, and it is also called a *multimeter*.

For the ohmmeter of Fig. 68, the value of $R1$ required to set the meter to zero when the battery voltage is exactly 1.5 volts is 10,000 ohms. The internal resistance of the milliammeter is $250/0.05 = 5000$ ohms. With the methods of Chapter 3, we can determine the equivalent parallel resistance of these two elements. The separate conductances are: adjustable shunt, $1/10{,}000 = 0.1$ millimho, where 1 millimho means $\frac{1}{1000}$ of 1 mho; milliammeter, $\frac{1}{5000} = 0.2$ millimho. The sum of the separate conductances is: $0.1 + 0.2 = 0.3$ millimho; and the equivalent resistance is $1/0.0003 = 3330$ ohms. This value is then the value of resistance in series with $R2$ and the battery when the test leads are shorted.

The total series resistance, $3330 + 16{,}670 = 20{,}000$ ohms, is known as the total internal resistance of the ohmmeter, and it determines the *half-scale* marking of the instrument. Reference to Fig. 68 shows that the half-scale value is 200, which when multiplied by 100 (one of the factors given on the range switch below the meter movement) gives the half-scale value of 20,000 ohms. If the unknown resistance R is exactly 20,000 ohms, attaching the test leads across R creates a series circuit having 20,000 ohms in R placed in series with 20,000 ohms inside the ohmmeter. The battery is then connected in series with $20{,}000 + 20{,}000 = 40{,}000$ ohms, and the battery current is only half as much as it was when the test leads were shorted together. If resistor $R1$ has not been changed since the adjustment with the leads shorted, the same fraction of the battery current will be diverted into the meter movement as before. Therefore, if the battery current is only half as large with $R = 20{,}000$ ohms as with the leads at x and y shorted, the meter current will be only half as large also. Since the meter had full-scale deflection for shorted leads, half this current will bring the pointer to the middle of the scale on the milliammeter. In summary, when the external unknown resistance R has the same size as the total internal resistance of the circuit of the ohmmeter, that particular value of R is known as the half-scale value for the ohmmeter. Other points can then be calculated to produce the nonuniform scale seen in Fig. 68.

Adding internal shunts to the circuit of the ohmmeter can produce lower values for the half-scale resistance, although the zero-adjusting resistor $R1$ may have to be reset after the range switch is changed. Adding series resistors to the circuit of the ohmmeter can produce higher half-scale values, but additional battery voltage must be provided. For the ohmmeter of Fig. 68, switching the ohms multiplier to $\times 1K$ (since K is an abbreviation from the metric system meaning 1000,

5: Meters and Instruments   75

Fig. 69. Multi-range volt-ohm-milliammeter. *The Triplett Electrical Instrument Co.*

$\times 1K = \times 1000$) introduces a 15-volt battery into the circuit to provide the higher voltage. Other arms on the same range switch also introduce the appropriate series resistance needed for the extra multiplication by 10. In Fig. 68, the slider resting under the 600-ma marking operates the range switch. Resistor $R1$ for zero-setting is operated by the small wheel recessed into the upper left corner of the case.

The simple circuit of Fig. 67 involves certain compromises in accuracy as the battery voltage drops with age. It is possible to bring the pointer back to zero with resistor $R1$ for a considerable drop in battery voltage. However, this change in $R1$ from 10,000 ohms also changes *slightly* the total equivalent internal resistance of the ohmmeter from 20,000 ohms (on the $\times 100$ range). With reduced battery voltage, the half-scale reading will no longer be exactly correct, and other readings

76 Essentials of Electricity

will be off somewhat in proportion. The addition of other elements to the basic circuit can minimize these changes. Figure 69 shows a more accurate multimeter that is equipped with such refinements, including a simple semiconductor circuit that protects the basic sensitive milliammeter in case of accidental misuse of the test leads.

Prob. 21-5. The resistance of a heating element is determined with the voltmeter-ammeter method of Fig. 66. When the voltmeter reads 14.3 volts, the ammeter reads 6.9 amperes. What is the resistance of the element?

Prob. 22-5. With the voltmeter-ammeter method of Fig. 66, the voltage across a large coil of fine magnet wire is found to be 80 volts for a current of 0.400 ampere. What is the resistance of the coil of wire?

Prob. 23-5. If the voltmeter current in Prob. 22-5 is estimated as 100 milliamperes and if the readings observed are unchanged, what is the actual value of the resistance of the wire?

Prob. 24-5. If the voltmeter reading of Prob. 22-5 is obtained on a range of 0–100 volts and if the figure-of-merit for the voltmeter is 1000 ohms per volt, estimate the voltmeter current.

Prob. 25-5. For the ohmmeter of Fig. 68 on the Ohms $\times$ 100 range, the value of $R1$ of Fig. 67 is 10,000 ohms for zero-set when the battery is at 1.5 volts, making the total internal resistance of the ohmmeter 20,000 ohms, as noted in this section. If the unknown resistance R is actually 5000 ohms, what value does the current in R have?

Prob. 26-5. For the zero-set conditions of the ohmmeter in Prob. 25-5 (with its test leads shorted), what is the current in meter M? What fraction of the battery current does this value represent?

Prob. 27-5. When unknown resistor R is being measured in Prob. 25-5, what is the current in meter M? What fraction of the battery current does this value represent?

8. Iron-Vane Instruments. A very sturdy construction for an ammeter or voltmeter is shown in Fig. 70. Inexpensive watch-case instruments often employ this design because it can stand extremely hard wear. Around the inside of the case extends a circular permanent magnet with poles N and S as marked. Directly between the magnetized poles is an oval piece of soft iron set in bearings, with a pointer attached. When there is no current in coil C, the lines of magnetic force through the iron bring the soft-iron oval to a horizontal position, holding the pointer at zero. With a current in coil C, the coil sets up a magnetomotive force that might, for example, set up flux lines directed downward out of coil C if the permanent magnet were not present. The combined effect of the current in coil C and of the perma-

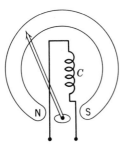

Fig. 70. Diagram of simple watch-case ammeter or voltmeter.

5: Meters and Instruments 77

nent magnet is to cause a distorted magnetic flux, with a shift of the axis of the lines from horizontal around in a clockwise direction. In keeping with the rules discussed in Chapter 4, the shifted flux lines turn the piece of iron around, moving the pointer upscale. Although the accuracy is not great, the mechanism is rugged. The coils of ammeters of this type have few turns of very heavy wire. Voltmeters have many turns of fine wire.

SUMMARY

An INSTRUMENT is a device that provides a measurement of a quantity immediately, at the moment of observation.

Electric instruments usually operate on the principle known as MOTOR ACTION.

The central part of an electric instrument is the PERMANENT-MAGNET MOVING-COIL device, often called a MILLIAMMETER.

The ranges of milliammeters can be extended with SHUNTS for AMMETERS, and with SERIES RESISTORS for VOLTMETERS.

The resistance necessary for a shunt of an ammeter is

$$\text{RESISTANCE OF SHUNT} = \frac{\text{RESISTANCE OF COIL}}{\text{FACTOR} - 1}$$

The series resistance to be used with a voltmeter is

$$\text{SERIES RESISTANCE} = \frac{\text{FULL-SCALE VOLTS}}{\text{FULL-SCALE COIL CURRENT}} - \text{COIL RESISTANCE}$$

Electric instruments that keep a permanent record of voltage or current are called RECORDERS.

A direct means for measuring resistance is called the VOLTMETER-AMMETER method.

OHMMETERS are self-contained instruments that provide a basic series circuit into which an unknown resistance can be inserted for quick measurement.

PROBLEMS

Prob. 28-5. A milliammeter gives full-scale deflection of 100 divisions when the coil current is 0.025 ampere and the voltage across its terminals is 0.200 volt. What is the resistance of the coil?

Prob. 29-5. A shunt is rated 0.050 volt when carrying 3 amperes. What is its resistance?

Prob. 30-5. A 50-millivolt millivoltmeter gives a full-scale deflection of 100 divisions when the coil current is 5 milliamperes. What is its coil resistance?

Prob. 31-5. If the movement of Prob. 30-5 is connected to the shunt of Prob. 29-5, what current in the external circuit does the 100-division full-scale deflection now represent?

Prob. 32-5. The instrument of Prob. 28-5 is to be provided with two ranges: 0–25 amperes and 0–250 amperes. What must be the resistances of the shunts?

Prob. 33-5. A voltmeter, with ranges of 0–7.5 volts and 0–150 volts, has a sensitivity of 100 ohms per volt on each range. What is the total resistance on each range?

Prob. 34-5. What current does the voltmeter of Prob. 33-5 require when measuring (*a*) 25 volts, (*b*) 90 volts, and (*c*) 150 volts?

Prob. 35-5. What current does the voltmeter of Prob. 33-5 require (for the lower range) when measuring (*a*) 2 volts, (*b*) 5 volts, and (*c*) 7.5 volts?

Prob. 36-5. The resistance of a large motor is to be measured by the voltmeter-ammeter method. When the current in the ammeter is 300 amperes, the voltmeter reads 2.4 volts on a range of 0–3 volts with a figure-of-merit of 10 ohms per volt. What is the resistance of the motor? Explain why the sensitivity of the voltmeter is not important for this measurement when the circuit of Fig. 66 is connected, the motor replacing lamp *L*.

Prob. 37-5. An ohmmeter has a battery supplying 3.0 volts and a total internal resistance, when set at zero, of 200 ohms. What is the current in the test leads when they are shorted together?

Prob. 38-5. When the test leads of the correctly zeroed ohmmeter of Prob. 37-5 are connected to an external resistance of 100 ohms, what is the current in the 100-ohm resistor?

Prob. 39-5. For Prob. 38-5, at what percentage of full-scale deflection does the pointer lie when the meter reads 100 ohms?

6
Series-Parallel Circuits

1. Circuit Connections. Almost all circuits that involve large quantities of electric energy are arranged so that the loads performing some useful function are connected in parallel. Although for a first estimate the resistances of the wires connecting these loads to their sources may be ignored, the actual size of the resistances is usually important enough that more careful calculations must include the connecting wires in the circuit. Therefore, most circuits are not either purely series circuits or purely parallel circuits; instead, most circuits are combinations of the two basic forms. One such combination in the preceding chapter was seen in the ohmmeter circuit of Fig. 67.

However, if many resistors and batteries are interconnected so that the form of their circuitry looks like a net used for fishing, it is possible to form an electric circuit that has no parts either in series or in parallel. Figure 71 shows an example, in a circuit called a resistance bridge, where middle arm M bridges points between the other resistors. If resistor M is removed, as in Fig. 72, the rest of the circuit can be classified as *series-parallel,* of which classification Figs. 7 and 8 were also examples.

In this chapter, only series-parallel circuits are considered. In addition to Ohm's law, two other laws are helpful for calculations on such circuits. When arrangements like the resistance bridge are encountered, the two *Kirchhoff's laws* (named for the man who first stated them) are absolutely essential for understanding the circuit behavior.

80 Essentials of Electricity

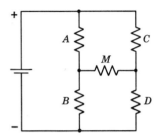

Fig. 71. A bridge circuit provides neither series nor parallel connections.

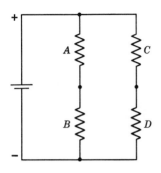

Fig. 72. A series-parallel connection.

2. Kirchhoff's Current Law. In Section 3-8, we reasoned that the current into a parallel circuit was the sum of the currents in the separate parallel branches. This idea can be extended to any terminal to which two or more electric-circuit elements have one end joined, in that electric current (like flowing water at a junction of filled pipes) must be accounted for completely in entering and leaving the terminal. Current does not build up or disappear at a terminal; the incoming current must be balanced by the same size of current leaving. If all the wires that bring current into a terminal are considered in one group and all the wires carrying current away from the terminal are considered in another group, *Kirchhoff's current law* may be stated as:

The sum of all the currents entering a terminal must be equal to the sum of all the currents leaving that terminal.

Figure 73 shows five elements joined together at a terminal, with the rest of the circuit omitted. The current that enters the terminal from the battery must all be accounted for in the sum of the currents leaving in the other branches.

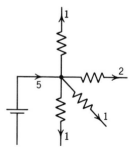

Fig. 73. The sum of the currents leaving the terminal is 5 amperes.

3. Kirchoff's Voltage Law. In Section 3-3 we saw that the work done along an entire series path was the sum of the work done in each part. Thus, the voltage across a series string of resistors is the sum of the separate voltages, usually equal to the voltage of the battery supplying the circuit. This idea may be extended to any closed path that we may trace in any electric circuit.

6: Series-Parallel Circuits   81

If we know the voltage between any two points in the path, we may consider that this voltage is essentially in parallel with the total voltage for the rest of the path. Thus, the voltage between any two points of a closed path is equal to the net voltage found across the rest of the parts of the path stretching between these points. The summation for the net voltage may be found for the parts of the path that lie on either side of the two points chosen.

However, some caution is necessary. Whereas it is usually simple to see which currents *enter* and which currents *leave* a terminal, it is not always as simple to see which voltages represent an *increase* in potential in some part of any path and which voltages represent a *decrease* in potential in some other part of the path. As an aid in describing these two ideas clearly, we say that an *increase* in potential is a *rise* in voltage and a *decrease* in potential is a *drop* in voltage. In these terms we can state *Kirchhoff's voltage law* as:

The sum of all the rises around any closed path of an electric circuit is equal to the sum of all the drops in voltage belonging in the same path.

Figure 74 repeats the form of Fig. 72, with the sizes of voltages added to show the summation of Kirchhoff's voltage law.

Example 1. Show some applications of Kirchhoff's voltage law in the circuit of Fig. 74.

For the battery potential of 12 volts, around the path including the battery and resistors A and B, we see that the 12-volt rise of the battery equals the sum of the drops, i.e., $(8 + 4)$ volts on resistors A and B. For the path clockwise around resistors B, A, C, and D, notice that the voltages on resistors A and B are *increases* or *rises* and that the voltages on resistors C and D are *drops*. Thus, for a clockwise tracing of the path B, A, C, D, the sum of the rises is $4 + 8 = 12$ volts, which must be the same as the sum of the drops in the remainder of the path: $3 + 9 = 12$ volts as a total drop across C and D.

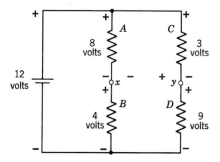

Fig. 74. Down any path from top to bottom the drop is 12 volts.

82   Essentials of Electricity

It is worthwhile to note that Kirchhoff's voltage law can be applied to any path that is closed, whether or not any current actually traverses that path. For example, note the letters x and y added to Fig. 74. For a clockwise closed path up through B to point x, across to point y, down through D, and thus back to B, there is clearly no current from x to y. However, we can still find the voltage between x and y. In the clockwise direction in which we are tracing, the voltage across B is a 4-volt *rise*, and there is a 9-volt *drop* across D. In order to satisfy Kirchhoff's voltage law, there must be a 5-volt *rise* from x going upward in potential to y. Then the sum of the rises in voltage equals the sum of the drops in voltage around this closed path. If a voltmeter of sufficiently high figure-of-merit (ohms per volt) is connected between points x and y, with its plus terminal connected to point y, it will read upscale to indicate 5 volts.

Recall from Chapter 2 that polarity for voltages is marked by a *pair* of signs. In Fig. 74 point x has different polarity designations, depending upon which pair is involved. There is no contradiction in this multiple marking. Point x is 8 volts negative with respect to the top of resistor A, and it is at the same time 4 volts positive with respect to the bottom or resistor B. Because in many circuits the important question concerns the *difference* in polarity and potential, the marking of a *pair* of polarity signs merely designates that most voltages are specified as the potential of one end of a circuit part relative to the other end.

Prob. 1-6. In Fig. 74, show how Kirchhoff's voltage law applies to the closed path taken counterclockwise from D through C, down through the battery, and over to D again.

Prob. 2-6. In Fig. 74, verify that there is a 5-volt drop in voltage from point y to point x, by considering the clockwise closed path from y to x, up through A and down through C, back to y.

Prob. 3-6. In Fig. 74, show how Kirchhoff's voltage law applies to the closed clockwise path defined as follows: start down through A to point x, go across to y, down through D, go left to the bottom of the battery, up through the battery back to A.

In Fig. 75, points p, q, s, and t have been added to the circuit of Fig. 74. Furthermore, currents have been specified as to magnitude and direction in resistors A and D. With this new information it is now possible to learn more about this circuit.

Example 2. What is the current in the battery of Fig. 75?

We note that each parallel branch of the circuit is in itself a simple series circuit. Since resistors A and B are in series, the current in resistor B (leaving terminal x) must be 2 amperes, the same as the current in resistor A (entering terminal x). Similarly, the current in resistor C is found to be 3 amperes. Another application of Kirchhoff's current law at terminal t shows that the battery current is 5 amperes (leaving terminal t), in order to be equal to the sum of the currents in resistors B and C ($2 + 3 = 5$ amperes entering terminal t).

Example 3. What is the equivalent resistance attached to the battery in Fig. 75?

6: Series-Parallel Circuits

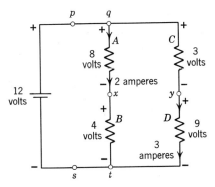

Fig. 75. *A* and *B* are in series; *C* and *D* are also in series.

Because we know the voltage of the battery, we know the voltage drop between points p and s is also 12 volts by Kirchhoff's voltage law applied in a clockwise direction around the left-hand mesh of the circuit. Knowing that the current from p to q is 5 amperes, we know that the equivalent resistance of the circuit must have 12 volts across it when the current through it is 5 amperes, if the battery current is to be the same with the equivalent resistance as with the original circuit. By Ohm's law, the equivalent resistance is $\frac{12}{5} = 2.4$ ohms. If this equivalent resistance is connected between p and s, in place of the original circuit as in Fig. 76, the current in at p and out at s and the voltage from p to s will have the same values as in Fig. 75: current, 5 amperes; voltage, 12 volts.

Prob. 4-6. In Fig. 75, what is the equivalent resistance of the series combination of resistors *A* and *B*?

Prob. 5-6. In Fig. 75, what is the equivalent resistance of the series combination of resistors *C* and *D*?

Prob. 6-6. What are the sizes of each of the resistors in Fig. 75?

Prob. 7-6. Combine the results of Probs. 4-6 and 5-6 to verify the value of 2.4 ohms for the equivalent resistance of the circuit.

4. Voltage Diagrams. Usually circuit diagrams are arranged on a page to be as neat and attractive as possible and to display the various parts with as few intricate twists of the wires as possible. Sometimes, how-

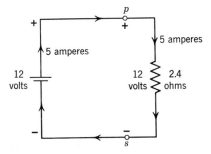

Fig. 76. Reduction of the circuit of Fig. 75 to a single equivalent resistance.

ever, it is helpful to rearrange a circuit diagram to emphasize some special way in which the circuit performs. One such helpful emphasis is the redrawing of a circuit to show as clearly as possible the various levels of voltage in the circuit.

The drawing known as a *voltage diagram* arrays the position of the various junctions or terminals with respect to some one chosen point. If there is just one generator or source in the circuit, it is common to choose its negative terminal as the fixed point. This point is then drawn at the bottom of the diagram, and the other points are arranged to stand higher to show that they have more positive, or higher, potential. Usually, the positive terminal of the source is shown at the top of the diagram, for it is the point of highest voltage when measured above the voltage of the negative terminal.

Example 4. Rearrange the circuit of Fig. 75 into a voltage diagram.

Since points s and t are at the same potential and this potential is that of the negative terminal of the battery, these points will be put at the bottom of the diagram. The next level higher in the diagram is point x, at 4 volts above the bottom. Point y is further above, at 9 volts rise from t. Finally at the top are points p and q, at the 12-volt point above the chosen *reference* or *datum* point. It is customary to call the datum level the level of 0 volts. The resulting voltage diagram is shown in Fig. 77, with the horizontal dashed lines added to emphasize the voltage levels.

There is no need to make the heights in a voltage diagram proportional to the actual voltages; labeling the diagram clearly takes care of the sizes. The visualization that the new spacing of the diagram gives makes the arrangement valuable. It is easy to see in Fig. 77 that point y is 5 volts more positive than point x.

Example 5. Redraw the series circuit of Fig. 11 (in Chapter 3) as a voltage diagram.

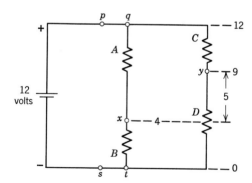

Fig. 77. A voltage diagram for the circuit of Fig. 75.

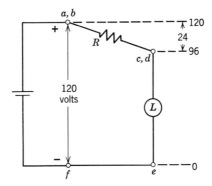

Fig. 78. A possible voltage diagram for the circuit of Fig. 11.

Although we do not know how the source establishes 120 volts, we can represent its effect by sketching a battery between points a and f. Let us choose point f as the reference or lowest point (0) for the diagram. We see, then, that Fig. 11 is already in the form of a voltage diagram. However, we can draw resistor R on a slanting line to represent its 24-volt drop from the highest point a to d and then draw the lamp vertically for the 96-volt drop remaining from the 120-volt total potential. Figure 78 shows this alternative form for a voltage diagram of the circuit of Fig. 11. The slanting section emphasizes the importance of the 96-volt drop by giving the lamp a more prominent place in the drawing.

Although the voltage diagram is helpful in visualizing the distribution of voltage in a circuit, it is important to recognize that many circuit diagrams are *not* arranged this way. Simply looking at the arrangement of the diagram does not always yield the information about the voltage distribution. The various voltage levels must be determined from Ohm's law and the two Kirchhoff's laws. With the levels thus calculated, the voltage diagram may be constructed as an aid to understanding the results of the calculations.

Prob. 8-6. In the circuit of Fig. 12 (in Chapter 3), choose the negative terminal of the 220-volt source as reference. Arrange a voltage diagram for this circuit, with the lamps drawn vertically and the 40-ohm resistor shown slanting at the bottom of the diagram.

Prob. 9-6. Draw a voltage diagram for the correctly zero-set ohmmeter ($R1 = 10,000$ ohms) of Fig. 67 when the test leads connected to x and y are shorted together.

Prob. 10-6. Draw a voltage diagram for the correctly zero-set ohmmeter ($R1 = 10,000$ ohms) of Fig. 67 when the external resistance R has a value of 20,000 ohms.

5. Parallel Loads on a Line. As noted in the first section of this chapter, most loads are usually connected in parallel. However, when the resistances of the connecting wires are considered, a series-parallel circuit results for many loads, especially for groups of lamps. The voltage at the lamps will be less than the source voltage when the drop in voltage in the line wires is accounted for.

The resistance of lamps differs from one lamp to another, even when they are all manufactured by the same company. After installation, even the voltage across each lamp is usually not exactly the same. For convenience in calculation, however, each lamp of a parallel group is assumed to require the same current. The error introduced by this assumption is usually too small to demand further study. With a little practice, the application of Ohm's law and Kirchhoff's laws will make relatively short work of solving even difficult arrangements of lighting loads.

Example 6. In a lighting system having four lamps, the circuit for which is shown in Fig. 79, each lamp requires 0.5 ampere at 110 volts. Find the generator voltage needed to supply the necessary line current while maintaining a load voltage of 110 volts at the lamps.

Consider the paralleled lamps as a single load. The current entering the lamp grouping at b must be $4 \times 0.5 = 2.0$ amperes, in accordance with Kirchhoff's current law. The lamp voltage is to be maintained at 110 volts. Therefore, the equivalent resistance of the lamps is $110/2.0 = 55$ ohms. We see that this form of problem with line wires having definite resistance and with parallel loads can be thought of as an equivalent series problem.

As a series circuit, the current in all parts must be the same; in this circuit, there would be 2.0 amperes in line wire ab, in the equivalent 55-ohm resistance, and in line wire cd. Since we know the resistances of the line wires, we can compute the voltage drop across each section as follows: (volts = ohms × amperes, by Ohm's law) for wire ab, $0.4 \times 2.0 = 0.8$ volt; for wire cd, also 0.8 volt. Adding all the voltages together (Kirchhoff's voltage law) as in a series circuit, we obtain the generator voltage between points a and d:

voltage across ab	= 0.8 volt
voltage across lamps	= 110. volts
voltage across cd	= 0.8 volt
total voltage across ad	= 111.6 volts

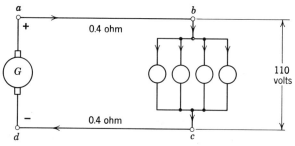

Fig. 79. The current in the line wires is the sum of the currents in the lamps.

The total voltage of 1.6 volts for the two line wires is not available for the lamps but it must be supplied by the generator in addition to the 110 volts needed for the lamps, if the lamps are to have a potential of 110 volts maintained at their terminals. Since this 1.6-volt potential is not available to the lamps, this total voltage lost in the line wires (going and returning) is called the *line drop,* or the *volts lost in the line.* Therefore, the answer to the problem is: the generator voltage needed is 111.6 volts.

Figure 80 shows a voltage diagram for these values. In addition, the various currents in the circuit are shown. Note that resistance symbols for the lines were not given in the original circuit of Fig. 79. It is often helpful to show these explicitly when constructing the voltage diagram. It also helps to construct the voltage diagram while the calculations are being carried on. Each current and voltage can be added to the complete circuit diagram as soon as it is found by computation.

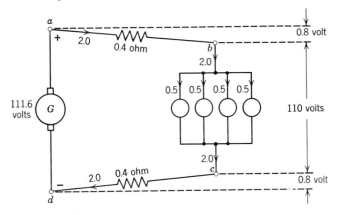

Fig. 80. A voltage diagram for the circuit of Fig. 79.

To summarize the *general method of attack* for this type of problem, note that there are *three* main steps:

First: Use Kirchhoff's current law to determine the current in each part of the circuit.
Second: Apply Ohm's law to find any unknown voltages in the circuit.
Third: Use Kirchhoff's voltage law to combine the voltages to obtain the total voltage.

If desired, construct a voltage diagram to show the voltage drops clearly. The line drop is the sum of the drops in each line wire, for this group of parallel lamps supplied over two line wires. Even without a voltage diagram, a circuit diagram that is labeled with each current as soon as it is computed, and with each voltage calculated in turn, offers great help in keeping the numberwork progressing smoothly.

88   Essentials of Electricity

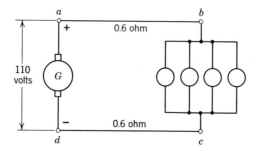

Fig. 81. The potential across the lamps is less than 110 volts.

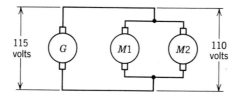

Fig. 82. Motors $M1$ and $M2$ are in parallel.

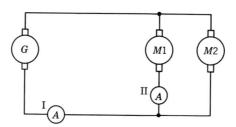

Fig. 83. Ammeter I shows current in the generator; ammeter II shows current in motor $M1$.

Prob. 11-6. The generator in Fig. 81 provides a terminal potential of 110 volts.
(a) If each lamp requires 1 ampere, what will the voltage at the lamps be?
(b) What will be the line drop?

Prob. 12-6. In Fig. 82, motor $M1$ requires 25 amperes and motor $M2$ requires 30 amperes. How much resistance can each line wire have if the generator voltage is 115 volts and if the motors are to be supplied with a potential of 110 volts at the end of the line wires?

Prob. 13-6. In Fig. 83, the line wires have a resistance of 0.15 ohm each, and the resistance of each ammeter can be considered to be zero. Ammeter I reads 60 amperes, and ammeter II reads 25 amperes. What is the current in motor $M2$? If the voltage at the motors is to be 110 volts, what must the terminal potential of the generator be?

Prob. 14-6. A potential of 110 volts is maintained at the distribution panel of an apartment house. One of the branch circuits from the panel has a total line resistance of 0.3 ohm. If a 6-ampere electric iron and a 5-ampere heater are both plugged into the end of this branch circuit at the same time, what will be the voltage across the appliances?

6. Ladder Groupings. In Example 6 and in the preceding problems, it will be noted that all parts of the line wires carry the same current. In practice, however, loads are not usually grouped all in parallel at the end of the line wires. Rather, systems of loads are usually arranged so that there are several *sections* of line wires, each section generally carrying different currents, the section nearest the generator carrying more than the sections further away when there is only one generator. A convenient name for describing such an arrangement is to call the system a *ladder circuit.* The circuit has the same form as a ladder resting on one side: the sections of line wires considered as two long lines are analogous to the sidepieces of the ladder, and the several groupings of loads running between these "sidepieces" are analogous to the rungs of the ladder.

A simple ladder distribution system for two lamps is shown in Fig. 84. Lamp A, requiring 3 amperes, is much nearer the generator than lamp B, requiring only 2 amperes. For the section of line between lamps A and B, the current is 2 amperes, the current required by lamp B. At point x, the current entering must be 5 amperes, by Kirchhoff's current law. Therefore the current in the line section between the generator and lamp A must carry the sum of the two load currents, or 5 amperes. There is a division of current at point x: of the 5 amperes entering point x, 3 amperes go to supply lamp A, and the remaining 2 amperes leave point x down the upper line wire to lamp B. Application of Kirchhoff's current law to point y shows that the entering currents, 3 amperes from lamp A and 2 amperes from the line section from lamp B, combine to make the total of 5 amperes for the lower line wire returning to the generator from lamp A.

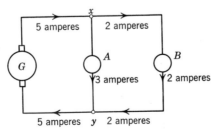

Fig. 84. Current distribution in a simple ladder circuit.

90   Essentials of Electricity

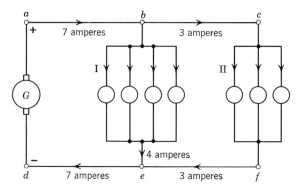

Fig. 85. Current distribution in a ladder circuit with lamp banks as "rungs."

In Fig. 85, there are two groups of paralleled lamps forming the ladder distribution. Since Group I requires 4 amperes and Group II requires 3 amperes, the current in line wire bc must be 3 amperes and the current in line wire ab 7 amperes, in accordance with Kirchhoff's current law. Similar applications of the current law at points e and f will verify the current distribution in the lower line wires fe and ed.

It is very often convenient to regard the current distribution of a ladder circuit as follows:

Begin with the load group farthest from the generator. Either think of the connecting line wires and this farthest load group as a series circuit, or apply Kirchhoff's current law at each terminal of the load group, to see that the current in that last line section is the same as the total load current of the farthest grouping. In Fig. 85, Group II requires 3 amperes, and thus the currents in line wires bc and fe must each be 3 amperes.

Proceeding back toward the generator, note that the next load grouping adds its current to the line current of the last section, to yield the current in the section of the line between the generator and the first load grouping, in accordance with Kirchhoff's current law at each terminal of the first load grouping. In Fig. 85, Group I requires 4 amperes. Line wire ab, therefore, must carry a current of 7 amperes to satisfy the current law at point b. Similarly, at point e, the separate currents entering this terminal add to yield 7 amperes in line wire ed.

In summary, when the load currents are known, the current distribution in the various parts of a ladder circuit can be found by working backward toward the generator from the load grouping farthest away. Successive applications of Kirchhoff's current law will yield the desired summations in each section of the line.

6: Series-Parallel Circuits 91

Prob. 15-6. If each lamp in Fig. 86 requires 0.3 ampere, how much current is there in each of the line wires: *ab, bc, ed,* and *fe*?

Prob. 16-6. In Fig. 87, each lamp requires 0.25 ampere. What are the values of line current in *ab, bf, dc,* and *ce*?

Prob. 17-6. In Fig. 88, the motors of three subway cars are at different distances from the generator because the cars are located at different points along the subway track. Arrowheads indicate the sliding electric contacts that connect the cars to electric power. The motor currents are: car 1, 50 amperes; car 2, 60 amperes; car 3, 70 amperes. In the third rail feeding power to the cars, what is the current in the third-rail section between: (*a*) cars 2 and 3? (*b*) cars 1 and 2? (*c*) the generator and car 1?

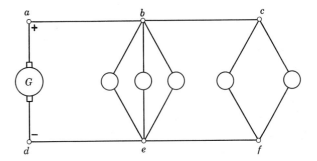

Fig. 86. The current is different in different sections of the line.

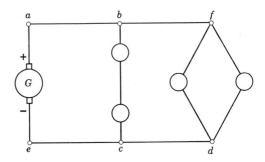

Fig. 87. Series and parallel groups of lamps supplied by generator.

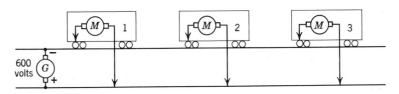

Fig. 88. Subway cars at different distances from power source.

92   Essentials of Electricity

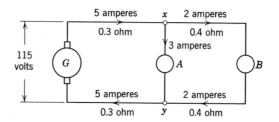

Fig. 89.   Line drop lowers the lamp voltages below 115 volts.

7. Line Drop for Fixed-Current Loads. If to the circuit of Fig. 84 are added the resistances of the line wires, the circuit may be drawn as in Fig. 89, which also shows the generator voltage. The resistance of each line wire between generator G and lamp A is 0.3 ohm. For each of the line wires between lamps A and B, the resistance is 0.4 ohm. Now the line drops can be found, and the voltages that result at each of the lamps can also be calculated. It is possible to apply the same three steps to this problem as were outlined in Section 6-5.

First Step.   Current Distribution from Kirchhoff's Current Law. The distribution for currents in this ladder circuit was already noted in the discussion of Fig. 84. If each lamp still requires the same current as it had before, the distribution of current is not changed.

Second Step.   Line Drop Calculations with Ohm's Law. When the actual resistance of a line wire is considered, a small voltage is required to set up the desired current in the line. For a current of 5 amperes from the generator to point x through the 0.3-ohm resistance of the line wire, Ohm's law requires a potential drop of

$$5 \times 0.3 = 1.5 \text{ volts (volts = amperes} \times \text{ohms)}$$

The current of 5 amperes in the other line wire from point y back to the generator also requires 1.5 volts as a drop in potential. Therefore, the line drop for the first section between the generator and lamp A is the sum of these two drops, or $1.5 + 1.5 = 3$ volts.

A similar calculation will give the line drop in the section between lamps A and B. For the upper wire, the drop is $2 \times 0.4 = 0.8$ volt. Since the lower wire has the same current and resistance values, its drop is also 0.8 volt. Then the sum of the two drops gives the line drop as 1.6 volts in the section between lamps A and B.

Third Step.   Load voltages from Kirchhoff's Voltage Law. Trace a clockwise path up through the generator, over to point x, down through lamp A to point y, and back to the generator. For this path, the generator rise in potential must be equal to the sum of the remaining voltages as drops. We know the line drop (3 volts) from the generator to lamp A, and we know the generator rise (115 volts). Therefore, we must *subtract* the line drop from the generator rise to obtain the voltage across lamp A:

$$115 - 3 = 112 \text{ volts}$$

A voltmeter placed across the generator terminals would read 115 volts. Because of the drop in the line, a voltmeter placed across lamp A would read only 112 volts. The remaining potential is dropped or lost along the line wires, and this 3-volt value is thus the line drop. Although wires of lower resistance could be installed to minimize the line drop, the additional cost of more nearly perfect wires may not be practical. Usually a compromise between the size of the line drop and the cost of a more expensive installation must be reached.

To find the voltage at lamp B, trace a path in the circuit going clockwise up through lamp A to point x, across to lamp B, down through lamp B, back to point y and lamp A again. In this path, Kirchhoff's voltage law requires that the 112-volt rise encountered going up through lamp A must be balanced by the sum of the remaining voltages as drops around the path. Since we have already found the line drop to be 1.6 volts for the section between lamps A and B, we again subtract the line drop from the known rise to find the unknown drop at lamp B:

$$112 - 1.6 = 110.4 \text{ volts}$$

A voltmeter connected across lamp B would read only 110.4 volts, not 112 volts, because of the additional line drop between the two lamps.

There is an additional path that might be traced for finding the voltage at lamp B. Go clockwise around the outside edge of the entire circuit, up through generator G, across to point x and on to lamp B, down through lamp B, back to point y and to the starting point at the generator. In this path we know the rise in voltage at the generator, and we can subtract all the known drops in voltage through the various line wires. Taken in order, the subtraction reads:

$$115 - 1.5 - 0.8 - 0.8 - 1.5 = 110.4 \text{ volts}$$

Thus we can confirm the value of voltage across lamp B previously obtained.

A voltage diagram of this circuit has been constructed in Fig. 90. The parts have been arranged to emphasize the fact that the line wires between lamps A and B may be considered to form a series circuit with lamp B. Then this series circuit is seen to form a parallel combination with lamp A. This arrangement stresses the idea that the line drop of

94 Essentials of Electricity

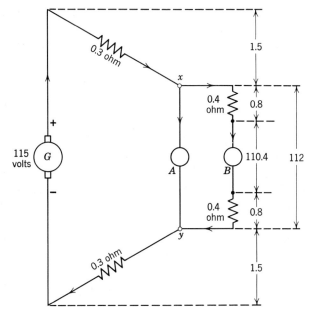

Fig. 90. A voltage diagram for the ladder circuit of Fig. 89.

1.6 volts between lamps A and B should be subtracted from the voltage at lamp A to get the voltage at lamp B.

8. Source Voltage for Fixed-Current Loads. Sometimes the voltage of the generator must be calculated when a given voltage is to be maintained across one set of lamps or other load at a distance from the generator. This calculation is merely a variation of the problem outlined in the preceding section, and it may be carried out with the same three basic steps used there.

Figures 85 and 91 are identical except that in Fig. 91 the resistance of each line wire is known, and the voltage across the Group II lamps is to be kept at 110 volts. To be calculated are the voltage across the generator and the voltage across the Group I lamps.

First Step. Current Distribution. Since each lamp requires 1 ampere, the current distribution is the same as in Fig. 85, and it is repeated in Fig. 91. It is essential to note that in all problems of this type the starting point is finding the current distribution from Kirchhoff's current law.

Second Step. Line Drop. In the line section between Groups I and II, the 0.4-ohm line wire bc carries 3 amperes. By Ohm's law there must be a drop across this 0.4-ohm resistance of $0.4 \times 3 = 1.2$ volts. Since there is a similar drop in line wire fe, the line drop in this section is $1.2 + 1.2 = 2.4$ volts.

For the line section between Group I and the generator, the line drop is found in a similar manner to be 2 × (0.2 × 7) = 2 × 1.4 = 2.8 volts.

Third Step. Kirchhoff's Voltage Calculations. For a clockwise path traced around the line section between Groups I and II, the rise in voltage from point *e* to point *b* (the desired voltage across Group I) is equal to the sum of the remaining drops in voltage in the path. Tabulated, these drops are:

voltage across *bc* = 1.2 volts
voltage across Group II = 110. volts
voltage across *fe* = 1.2 volts
voltage across *be* (sum) = 112.4 volts

Stated another way, the voltage across Group I is equal to the voltage at Group II (110 volts) plus the drop (2.4 volts) in the line section between Groups I and II, or 110 + 2.4 = 112.4 volts.

Similarly, in a clockwise path drawn around the line section between the generator and Group I, the generator voltage must be equal to the Group I voltage plus the line drop of the intervening section. Therefore the generator voltage is 112.4 + 2.8 = 115.2 volts. Although we may show the successive drop in voltage along the system by a voltage diagram, we may also present the same information in tabular form:

voltage across generator *ad* = 115.2 volts
voltage across Group I *be* = 112.4 volts
voltage across Group II *cf* = 110. volts

For the various ladder problems with fixed-current loads, it should be clear that the same method has been followed. This method is again outlined here for review.

First Step. Working with Kirchhoff's current law, mark the current distribution on a diagram, beginning at the section farthest from the generator.

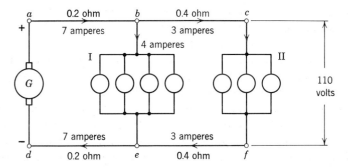

Fig. 91. Voltage at Group II must be 110 volts.

96 Essentials of Electricity

Second Step. Compute the line drop in the different sections of the line, with the aid of Ohm's law.

Third Step. With Kirchhoff's voltage law, or with application of its simpler forms for series and parallel circuits, compute the voltages that remain unknown in the system.

Prob. 18-6. For each lamp in Fig. 92 the current is 2 amperes. Motor M requires 15 amperes. Find: (a) the line drop for each section of the line, (b) the voltage across the lamps, and (c) the voltage across the motor.

Prob. 19-6. What is the resistance of each lamp in Prob. 18-6?

Prob. 20-6. In Fig. 86 of Prob. 15-6:

resistance of ab = resistance of de = 1.5 ohms
resistance of bc = resistance of ef = 2.0 ohms
voltage across be = 110 volts

Each lamp requires a current of 0.75 ampere. Find: (a) the line drop in each section of the line, (b) the voltage across ad, and (c) the voltage across cf.

Prob. 21-6. Find the resistance of each lamp in Prob. 20-6.

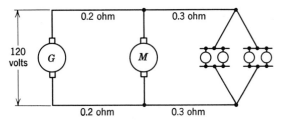

Fig. 92. Motor and lamps operating at different voltages.

9. Ladder-Circuit Voltages for Fixed Load Resistance. Some ladder circuits have slightly different features from the series-parallel distribution schemes just considered. In automotive and electronic wiring, the reference point is very often a common connection for many parts of the circuit, either to a metal frame or chassis, or to a sufficiently heavy conductor or *bus bar* whose resistance can be completely ignored in most calculations. Sometimes an important function of the ladder circuit is to produce a definite and large reduction in voltage to the load farthest from the source. Then actual resistors are introduced into the upper "sidepiece" of the ladder, and the bus bar or common point with essentially zero resistance is the lower "sidepiece."

In these modified ladder circuits, the loads may have fixed values of resistance, rather than a fixed current. When all the resistances of such a circuit are known, the problem of maintaining a definite voltage at the load farthest from the source requires slight modification of the basic three-step method.

Example 7. All the resistance values are stated in ohms for the ladder circuit of Fig. 93. The potential desired across the 30-ohm load is 18 volts. Find all the voltages and currents in the remainder of the circuit.

Although the solution of this problem proceeds with the same basic three steps involving Ohm's law and the two Kirchhoff's laws, it is necessary to apply the three steps to each mesh or space between the ladder "rungs" in sequence, starting with the last mesh at the 30-ohm load. Here, the current in the farthest load is not known, as it was in the previous problems, but it must first be computed from Ohm's law.

Since the source battery is shown with its + terminal at the top of the diagram, it is reasonable to decide that all of the load currents are directed downward. In the 30-ohm load, the current directed downward is $\frac{18}{30} = 0.6$ ampere. From the fact that the 10-ohm and 30-ohm resistors are in series, it follows that the current in the 10-ohm arm is also 0.6 ampere. Note, however, that this series current distribution is as far as we can carry the first-step process of finding the entire current distribution at this point in the calculations.

Now we can apply the second step to find the voltage drop along the section of path cd from Ohm's law. The drop in the 10-ohm resistor is $10 \times 0.6 = 6$

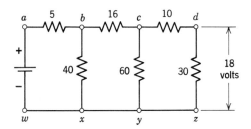

Fig. 93. Voltage across the 30-ohm load must be 18 volts.

volts. Since we do not know any of the remaining currents in the entire current distribution yet, we cannot find any more of the drops in the "sidepieces" of the ladder (sometimes called the *series arms* of the ladder; the rungs are called *shunt arms*). We proceed to the third step of applying Kirchhoff's voltage law to the path around the last mesh, $ycdzy$.

Because of the ideal (or practically ideal) connecting wire between points z and y, the 0.6-ampere current through this part of the path produces no voltage; in other words, points z and y are at the same potential. Therefore, for this modified ladder network, the voltage drop from c to y equals the sum of the drops across cd and dx. Thus the voltage across the 60-ohm resistor is $6 + 18 = 24$ volts. The results of these computations are shown in Fig. 94 for the right-hand mesh of this circuit. Polarity marks have been added to the voltage designations to show clearly the summation process of Kirchhoff's voltage law. The polarity pairs for each resistor show a drop in voltage (from + to −) in the same direction as the current in that resistor, as noted in Chapter 2.

Now we can start the procedure over again. The current directed downward in the 60-ohm resistor is, by Ohm's law, $\frac{24}{60} = 0.4$ ampere. At point c, we apply Kirchhoff's current law to learn more about the current distribution of the cir-

98 Essentials of Electricity

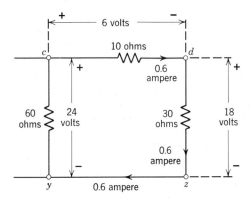

Fig. 94. Analysis of right-hand mesh of Fig. 93.

cuit. The current entering point c equals the sum of the currents leaving, or the current in the 16-ohm resistor of Fig. 93 is $0.4 + 0.6 = 1$ ampere. The same summation occurs at point y to give the current from y to x as 1 ampere also.

Ohm's law yields the drop in the 16-ohm resistor as $1 \times 16 = 16$ volts. From the voltage law applied around the middle mesh of this ladder circuit, the drop across the 40-ohm resistor is equal to the sum of the voltage across bc and cy, or $16 + 24 = 40$ volts. Since the wires connecting all the lower points of the circuit (w, x, y, z) have negligible resistance, each of the lower arms of this ladder circuit has zero voltage as its drop. At this point in the calculations, the process in continuing the numberwork to the left of points bx repeats exactly in the same fashion as the process begun at the 60-ohm resistor.

The current directed downward in the 40-ohm resistor is 40 volts/40 ohms = 1 ampere. Combining the currents leaving junction b gives the current in the 5-ohm resistor as $1 + 1 = 2$ amperes. The drop across the 5-ohm resistor is $2 \times 5 = 10$ volts, from Ohm's law. Finally, the battery voltage is the sum of the voltage across ab and bx, or $10 + 40 = 50$ volts.

A complete current distribution for this circuit is shown in Fig. 95. For ease in visualizing the voltage drops, Fig. 96 presents the results of the calculations in a voltage diagram.

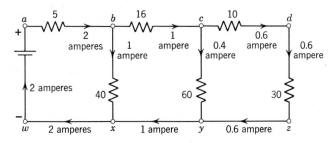

Fig. 95. Current distribution for the ladder circuit of Fig. 93.

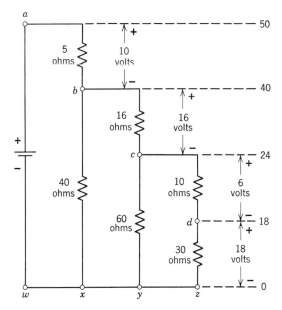

Fig. 96. A voltage diagram for the circuit of Fig. 93.

10. Ladder-Circuit Voltages with Fixed Source Voltage. In many circuits having the form of Fig. 93, all the resistance values are known, but the voltage specified is the source voltage. Since it is not clear how the current distribution for such a network can be obtained by starting with the current in any one resistor, a further modification of the basic three-step method is needed.

In Chapter 3, methods of combining resistors in series and resistors in parallel were considered. The possibility of determining an *equivalent resistance* has been exploited in Section 5-7, concerning the ohmmeter. Such a process of calculating an equivalent resistance for a circuit is known as *network reduction*. When two or more resistances of a circuit are combined into a single equivalent resistance, the original circuit is reduced somewhat in size. When all the resistances have been combined until only one resistance remains connected to the single source, that one resistance is said to be equivalent to the original circuit or network of resistances, inasmuch as the equivalent resistance takes the same current from the source as the original circuit. Example 3 of this chapter showed the process of network reduction to a single equivalent resistance for the circuit of Fig. 75.

Therefore, when the currents and voltages are to be calculated for a ladder network of known resistances when the source voltage is speci-

100   Essentials of Electricity

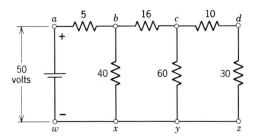

Fig. 97. The source voltage is specified for this circuit.

fied, one good approach to the problem is to apply network reduction. The ladder of resistances is reduced to a single equivalent resistance through series-parallel combinations of resistors. The source current is calculated from Ohm's law, dividing the source voltage by the equivalent resistance. Once the source current is known, the entire network reduction may be unfolded again, step by step, to obtain all the voltages and currents in the original circuit.

Example 8. The circuit of Fig. 93 is repeated in Fig. 97, except that in Fig. 97 the source voltage is specified as 50 volts. Determine all the circuit voltages and currents.

Since all the resistances have the same values as in Fig. 93 and since the applied voltage is the same as that found in Example 7, the results are already tabulated in Figs. 95 and 96. However, the order in which these results are calculated is quite different for this problem.

First apply network reduction to Fig. 97. Combine the 10-ohm and 30-ohm resistors in series to get an equivalent resistance of 40 ohms. This 40-ohm equivalent resistance connects between points c and z of Fig. 97, in place of the series combination. Thus, the 60-ohm resistance between points c and y is in parallel with the equivalent 40-ohm resistance between points c and z, as shown in Fig. 98.

For the parallel reduction, the conductances are: 60 ohms, $\frac{1}{60} = 16.67$ millimhos; 40 ohms, $\frac{1}{40} = 25$ millimhos. The sum of the two conductances gives the total conductance between points c and y as $16.67 + 25 = 41.67$ millimhos. Since the equivalent resistance for a conductance is the reciprocal of the conductance, the resistance equivalent to the parallel combination between points c and y is $1/0.04167 = 24$ ohms. The resultant reduction is shown in Fig. 99.

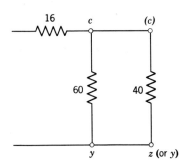

Fig. 98. Network reduction applied to the right-hand series elements of the circuit of Fig. 97.

6: **Series-Parallel Circuits** 101

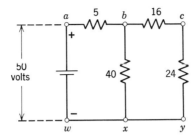

Fig. 99. Result of network reduction applied to the right-hand mesh of Fig. 97.

Note that Fig. 99 has the same form of its right-hand mesh as the mesh in Fig. 97, even though the sizes of resistors are different. Thus the same process of calculation is repeated to bring about further reduction. Add the 16-ohm series arm to the 24-ohm equivalent shunt arm to get another equivalent resistance of 40 ohms between points b and y. Note that this 40-ohm equivalent resistance is in parallel with the original 40-ohm shunt arm of Fig. 97. For two equal resistances, the equivalent parallel resistance (from Chapter 3) is half of one of them, i.e., the equivalent resistance between points b and x is $\frac{40}{2} = 20$ ohms. Last, add the 20-ohm equivalent shunt arm to the original 5-ohm series arm to get the single equivalent resistance for Fig. 97, namely, $5 + 20 = 25$ ohms, as shown in Fig. 100.

Since the voltage of the battery is specified as 50 volts in Fig. 97, the current into the equivalent resistance can be found with Ohm's law to be $\frac{50}{25} = 2$ amperes. This value must also be the current in the battery, in Fig. 100 to satisfy Kirchhoff's current law, and in Fig. 97 because the two diagrams are equivalent with respect to the battery terminals.

Then the process of unfolding the network reduction can begin. With a current of 2 amperes in the original 5-ohm resistance, there must be a drop of 10 volts from point a to point b of Fig. 97. Kirchhoff's voltage law applied to the source mesh of Fig. 97 yields $50 - 10 = 40$ volts as the drop from point b to point x. With

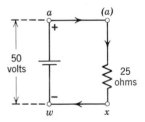

Fig. 100. Reduction of the circuit of Fig. 97 to a single equivalent resistance.

a drop of 40 volts across the original 40-ohm resistor, the current downward through it must be 1 ampere. These results are shown in Fig. 101.

At point b, there is a current of 2 amperes entering; the current is 1 ampere leaving downward through the 40-ohm shunt arm. According to Kirchhoff's current law, there must be a current of 1 ampere leaving from point b through the 16-ohm series arm of Fig. 97 to yield the balance: 2 amperes entering $= (1 + 1)$ amperes leaving. Knowing this value of 1 ampere in the 16-ohm resistor permits repeating another cycle of the calculations to unfold the network reduction.

The drop across the 16-ohm resistor is $16 \times 1 = 16$ volts from b to c. Application of Kirchhoff's voltage law clockwise from x to b to c to y of Fig. 97 gives the drop across the 60-ohm shunt branch as: (40 volts, rise from x to b) —

102 Essentials of Electricity

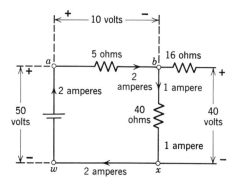

Fig. 101. Calculated values for the left-hand mesh of Fig. 97.

(16 volts, drop from b to c) = (24 volts, drop from c to y). With Ohm's law, the current downward in the 60-ohm resistor is $\frac{24}{60} = 0.4$ ampere. From Kirchhoff's current law at junction c, the current leaving c headed to the right through the original 10-ohm series arm of Fig. 97 is $(1 - 0.4) = 0.6$ ampere. This value is also the current in the 30-ohm resistor, since it is in series with the 10-ohm branch.

Furthermore, the voltage from c to d is $(10 \times 0.6) = 6$ volts. Applying the voltage law to the right-hand mesh of Fig. 97 gives $24 - 6 = 18$ volts across the 30-ohm load. This value may also be verified by Ohm's law: $30 \times 0.6 = 18$ volts. As noted above, all of these voltages and currents are the same as those in Example 7. The order of obtaining them is different because the source voltage was specified in Example 8, rather than the voltage of the last shunt arm as in Example 7. For a careful review, all of the calculations of Example 8 should be checked against the results shown in Figs. 95 and 96.

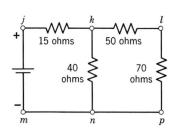

Fig. 102. Ladder circuit with fixed load voltage.

Prob. 22-6. In the ladder circuit of Fig. 102, the voltage across the 70-ohm load must be 14 volts. Find the source voltage required.

Prob. 23-6. The source voltage for the ladder circuit shown in Fig. 103 is 38 volts. Determine the current in the 50-ohm load.

Prob. 24-6. In Fig. 102, if the source voltage is specified as 250 volts (instead of the specification of load voltage made in Prob. 22-6), what must the new current value be in the 70-ohm load?

Prob. 25-6. If the source voltage in Fig. 103 is 500 volts, what is the current in the 12-ohm load?

Prob. 26-6. The circuit shown in Fig. 104 is known as a voltage-divider arrangement, supplying various voltage levels from one fixed source potential.

(a) Make a new drawing of this circuit to show the parts arranged in the form of a ladder circuit.

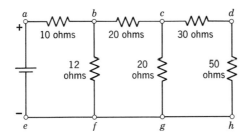

Fig. 103. The value of only one current is requested.

(*b*) Loads *A*, *B*, and *C* require 5 milliamperes (ma), 25 ma, and 50 ma, respectively, and load *B* requires a potential of 250 volts. Calculate the source voltage required.

Prob. 27-6. In the voltage-divider circuit of Fig. 104, let the source potential be 105 volts; let loads *A*, *B*, and *C* have fixed resistances of 12,000 ohms, 9000 ohms, and 6000 ohms, respectively. Calculate the voltage across load *C*.

Prob. 28-6. In the circuit of Fig. 104, let the load resistances be those of Prob. 27-6. Instead of specifying the source potential, choose the current in load *C* as 6 milliamperes. Calculate all the load voltages and the source potential.

11. The Unit-Ampere Method. For ladder circuits with known resistances and a specified source voltage, as in Fig. 97, there is another method available for finding all the currents and voltages. This method is based on the fact that each resistor must have a current in it that satisfies Ohm's law for the voltage across that resistor. Similarly, Ohm's law of proportionality must be satisfied for the equivalent resist-

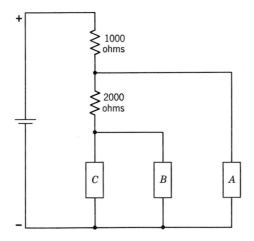

Fig. 104. A voltage-divider circuit.

104   Essentials of Electricity

ance of the network reduction. Therefore, if the battery voltage in Fig. 97 were doubled, i.e., if the source voltage were 100 volts instead of 50, the current in the equivalent resistance would be doubled also. With 4 amperes instead of 2 in the 5-ohm series arm, the voltage drop in the series arm would also be doubled to become 20 volts instead of 10 volts. Because of the proportionality represented by resistances, it follows that doubling the source voltage would double *every* voltage and current in the circuit if all the resistances kept the same values. Similarly, multiplying the original source voltage by some decimal multiplier such as 0.83 would cause *every* voltage and current to be multiplied by 0.83 as well.

Because of the relationship of proportionality established by Ohm's law, it is not necessary to carry through the process of network reduction for a ladder network when the source voltage is specified. Rather, any value of current may be guessed or assumed for the load farthest from the generator and then the complete problem may be solved as if this current (or the voltage across the load) had been specified, as outlined in Section 6-9. Since any effort at simplifying the numberwork is desirable, the customary choice is to assume 1 ampere for the load current, a choice that explains the name of this procedure: the 1-ampere method, or the *unit-ampere method*. When the solution is completed to the point where the source voltage has been calculated, as explained in Section 6-9, it will usually be found that the answer is not the same as the given source voltage. If the *given source voltage* is *divided by* the *calculated source voltage*, the decimal ratio obtained shows the fraction by which all the calculated answers differ from the true answers. To finish the problem, all the calculated answers are *multiplied* by the voltage fraction to get the true answers. Since this procedure involves working through the details of the circuit only once, the unit-ampere method usually saves time over the network-reduction scheme. To illustrate the method, the familiar problem of Example 8 will be worked again so that the true answers can be checked against the previous results.

Example 9. Compute all the voltages and currents for Fig. 97 by the unit-ampere method.

First assume that the current in the 30-ohm load is 1 ampere, directed downward. The current in the 10-ohm series arm will thus also be 1 ampere, and the voltage drops across *cd* and *dz* will be 10 volts and 30 volts, respectively. Adding these two voltages together by the voltage law gives the drop across the 60-ohm load as 40 volts. Ohm's law then yields the current in the 60-ohm load as $\frac{40}{60} = 0.667$ ampere.

Applying Kirchhoff's current law to junction *c* shows that the current in the 16-ohm series arm is $0.667 + 1 = 1.667$ ampere from *b* toward *c*. The drop in the series arm is $16 \times 1.667 = 26.67$ volts. Kirchhoff's voltage law combines the drops across bc and cy to give the voltage across the 40-ohm load: $26.67 + 40 = 66.67$ volts. From Ohm's law the current in the 40-ohm load is $66.67/40 = 1.667$ amperes.

Repeating the sequence of calculations just outlined completes the assumed solution. At junction b, the current entering must be $1.667 + 1.667 = 3.33$ amperes, with the last number rounded off. This value of 3.33 amperes is also the current in the source. The drop in the 5-ohm arm is $5 \times 3.33 = 16.67$ volts, again including the rounded-off decimals. The voltage law leads to the source potential as $16.67 + 66.67 = 83.33$ volts, rounded off, or ($2\frac{50}{3}$) volts if expressed as an improper fraction. All of these calculated values are placed on the circuit diagram of Fig. 105 for convenient reference.

The calculated source voltage of 83.33 volts is, of course, too high, since we know that the specified voltage is 50 volts. The fraction needed for obtaining the true answers is obtained by dividing the given source voltage by the calculated source voltage:

$$\frac{50}{83.33} = \frac{50}{(2\frac{50}{3})} = \frac{3}{5} = 0.6$$

Multiplying each of the calculated answers shown in Fig. 105 by 0.6 produces the set of true answers. For example, in the 5-ohm arm, the calculated current of the 1-ampere method is 3.33 or $\frac{10}{3}$ amperes. The true current is only 60% as large: $3.33 \times 0.6 = 2$ amperes, as found previously. Similarly, the true current in the 30-ohm load resistor is 0.6 ampere, and not 1 ampere as assumed. All of the answers, both for voltages and currents obtained by multiplying the answers of Fig. 105 by 0.6, should be checked against the sets of answers shown in Figs. 95 and 96.

There are some types of circuit problems that require network reduction for efficient solution. If the value of equivalent resistance at any point in the circuit must be known, network reduction is extremely important. However, the unit-ampere method avoids making detailed

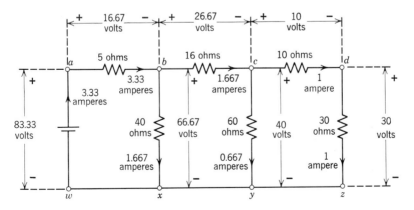

Fig. 105. For the circuit of Fig. 97, these values were obtained by the unit-ampere method.

106   Essentials of Electricity

calculations twice on a ladder circuit. This saving is worthwhile if all the voltages and currents are to be found, and it is especially valuable when the current is to be found only in the load farthest from the source.

Prob. 29-6. In Fig. 86, the resistances of the line wires are:

resistance of ab = resistance of de = 2.0 ohms
resistance of bc = resistance of ef = 3.0 ohms

If each lamp has a resistance of 90 ohms and if the source voltage is 120 volts, use the unit-ampere method to find the current in each of the lamps.

Prob. 30-6. In Fig. 87, each of the sections of line wire (ab, bf, ec, cd) has a resistance of 2 ohms and each lamp has a resistance of 80 ohms. For a generator voltage of 120 volts, use the 1-ampere method to find the voltage across each of the lamps in branch bc.

Prob. 31-6. Solve Prob. 23-6 by the unit-ampere method.

Prob. 32-6. Solve Prob. 24-6 by the unit-ampere method.

SUMMARY

Series-parallel circuits can be separated into small sections, and a basic three-step method can be applied to each section. METHOD. *First Step.* With the aid of Kirchhoff's current law, find the current distribution for as much of the circuit as possible.

Second Step. Compute, by means of Ohm's law, the voltages in all resistances in which currents are known.

Third Step. Apply Kirchhoff's voltage law to combine the voltages obtained as a result of the preceding steps.

For circuits with SPECIFIED LOAD CURRENTS, the basic method may be applied directly for a complete solution for all voltages and currents.

The order in which the basic steps are applied must be changed for circuits with fixed, SPECIFIED RESISTANCES.

Circuits arranged in the form of a ladder are called LADDER NETWORKS.

Modified application of the three-step method leads to solutions of ladder circuits with fixed resistances and ONE SPECIFIED LOAD VOLTAGE OR CURRENT.

Combination of the various series and parallel arrangements of resistances produces a single EQUIVALENT RESISTANCE attached to the original source.

The process of calculating an equivalent resistance is called NETWORK REDUCTION.

When a load current of 1 ampere is assumed in the most remote load in a ladder network, the resulting process of solution with a voltage fraction is called the UNIT-AMPERE METHOD.

Ladder circuits in which the source voltage and all resistances are specified may be solved either with NETWORK REDUCTION or with the UNIT-AMPERE METHOD.

PROBLEMS

Prob. 33-6. For Fig. 87 the line resistances are:

resistance of ab = resistance of ec = 0.8 ohm
resistance of bf = resistance of cd = 0.7 ohm

For a generator voltage of 125 volts, each lamp requires 1.5 amperes. Find: (*a*) the current in each section of the line, and (*b*) the line drop in each section of the line.

Prob. 34-6. In Prob. 33-6, find the voltage across the lamp groups, i.e., across *bc* and across *fd*.

Prob. 35-6. Assuming that the two lamps in the group across *bc* of Prob. 33-6 have the same resistance, find the resistance of each.

Prob. 36-6. In Prob. 33-6, what is the resistance of each lamp across *fd*?

Prob. 37-6. For the subway line shown in Fig. 88, the resistances of the third-rail sections are:

> between generator and car 1 = 0.5 ohm
> between car 1 and car 2 = 1.2 ohms
> between car 2 and car 3 = 0.4 ohm

The resistances of the track sections are:

> between generator and car 1 = 0.05 ohm
> between car 1 and car 2 = 0.1 ohm
> between car 2 and car 3 = 0.04 ohm

The currents required by cars 1, 2, and 3 are 50 amperes, 60 amperes, and 70 amperes, respectively. With the generator producing 600 volts, find: (*a*) the voltage drop in each section of the third rail, (*b*) the voltage drop in each section of the track, and (*c*) the voltage across the apparatus in each car.

Prob. 38-6. Assume that the voltage of the generator in Fig. 88 is not known, but that the voltage at car 2 is 550 volts. All other data are the same as in Prob. 37-6. Find: (*a*) the voltage drop in each section of the third rail, (*b*) the voltage drop in each section of the track, and (*c*) the voltage across the apparatus in each car.

Prob. 39-6. Each lamp in Fig. 106 requires 1.4 amperes. Find: (*a*) the amount and direction of the current in each of the following sections of the line wires: *ft, et, bc, ab*; and (*b*) the voltage drop in each of the four line-wire sections.

Prob. 40-6. Find the voltage across each lamp of Fig. 106 if the generator voltage is 125 volts and each lamp requires 1.4 amperes.

Prob. 41-6. Find the voltage across the generator and the Group I lamps in

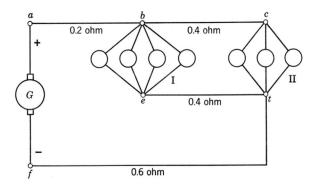

Fig. 106. Voltages at the lamps are nearly equal with this unusual arrangement.

108 Essentials of Electricity

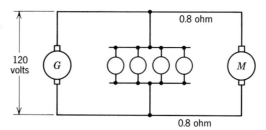

Fig. 107. Voltage at the generator is maintained at 120 volts.

Fig. 106 if the voltage across the Group II lamps is 112 volts and each lamp requires 1.4 amperes.

Prob. 42-6. In Fig. 107, each lamp requires 2 amperes and the motor current is 10 amperes. The voltage across the lamps must be 112 volts. Find: (a) the line drop in the wires between the lamps and the motor, (b) the voltage across the motor, (c) the resistance of the line between the generator and the lamps, and (d) the resistance of each lamp, if it is assumed that their resistances are all the same.

Prob. 43-6. In Fig. 107, assume that the resistance of the line wires from the generator to the lamps is 0.26 ohm each and that the voltage across the lamps is unknown. What will the voltage be across the motor when the lamps are not turned on?

Prob. 44-6. What will be the voltage across the lamps in Prob. 43-6 when the motor is not running?

Prob. 45-6. What will the voltage across the lamps in Prob. 43-6 become when the motor and the lamps are all operating?

Prob. 46-6. If the locations along the line of the motor and the lamps of Prob. 43-6 are interchanged, what will the new voltages be across the motor and across the lamps?

Prob. 47-6. At the generator's end of a short electric distribution line, the voltage is 120 volts. With a load of 200 amperes, the voltage at the consumer's end of the line is 116 volts. What is the resistance of each of the two line wires?

Prob. 48-6. On the consumer's end of the line in Prob. 47-6 is a motor. When the motor is started alone on the line, the voltage at the motor drops to 112 volts. When the motor gets up to its steady speed, the motor's current becomes smaller than the starting value and its terminal voltage becomes steady at 115 volts.

(a) What is the starting current of the motor?

(b) What is the running current of the motor?

Prob. 49-6. Fig. 108 shows a circuit found in some transistorized pocket radios for supplying the proper currents (called polarizing or biasing currents) to a transistor, for which values of equivalent resistance have been shown between points b, c, and e, (letters standing for base, collector, and emitter). The operation of the materials in the transistor is such that when the base current is 0.2 milliamperes, the collector current is 3 milliamperes, as shown in Fig. 108.

(a) What is the value of current directed into the emitter terminal?

(b) Find the emitter-to-base voltage, i.e., the drop from point e to point b, with the aid of the voltage law.

6: Series-Parallel Circuits

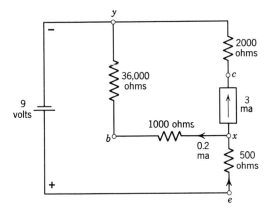

Fig. 108. Biasing circuit for a transistor.

(c) With the voltage law applied to the path through the battery from e to y, around from y to c to x, and back to e, find the voltage from collector to emitter and state whether the collector terminal is positive or negative with respect to the emitter terminal.

Prob. 50-6. When stacks of hinged graphite plates are mounted in a suitable holder, the assembled device is known as a *carbon-pile adjustable resistor*. It is sometimes inserted in series with the field winding of a motor or generator for smooth control of the current in that circuit. Two such carbon-pile resistors are available, each having a minimum resistance of 14 ohms and a maximum resistance of 50 ohms. Specify the range of resistance that may be covered if the resistors are connected: (*a*) in parallel and (*b*) in series.

Prob. 51-6. If three of the carbon-pile resistors described in Prob. 50-6 are available, specify all the ranges of resistance that can be covered with series and parallel combinations.

Prob. 52-6. Fig. 109 shows the connection of a rheostat R into the field circuit of a motor. This adjustable resistor must change the current smoothly without interrupting the field circuit while making possible a range in resistance from 23

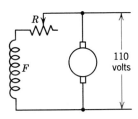

Fig. 109. Rheostat R controls the current in field F of a motor or generator.

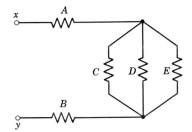

Fig. 110. Series-parallel connection of resistors.

110   Essentials of Electricity

ohms to 70 ohms. What one arrangement of the three carbon-pile resistors of Prob. 51-6 could be permanently connected to the motor's field circuit to satisfy these conditions?

Prob. 53-6. If the remainder of the field circuit in Fig. 109 has a resistance of 75 ohms, between what limits can the field current be adjusted with the resistance range specified in Prob. 52-6 if the terminal potential is maintained at 110 volts?

Prob. 54-6. The letters on the resistors of the circuit shown in Fig. 110 specify the resistance values as follows: $A = 1$ ohm, $B = 2$ ohms, $C = 3$ ohms, $D = 4$ ohms, and $E = 5$ ohms. What is the single resistance equivalent to the combination at terminals x and y?

Prob. 55-6. Solve Prob. 25-6 by the unit-ampere method.

Prob. 56-6. Solve Prob. 27-6 by the unit-ampere method.

7

Electric Power and Energy

1. The Time Rate of Doing Work. In many problems concerning electric circuits, there is great emphasis on the fact that Ohm's law is a relationship of proportionality. The number of ohms associated with an electric device states the *ratio* of voltage to current for that device. It is the idea of *ratio* that makes possible the statement that the current will double in a circuit of resistance when the applied voltage is doubled. Another important emphasis in the study of electric circuits is the special *ratio* that is formed when the divisor is time, usually in seconds. When the ratio is based on time, the resulting number is called a *time rate*. Familiar time rates are speeds of automobiles in miles per hour, the climbing rate of rockets in feet per second, and current in amperes—a unit that means coulombs per second past a point in the circuit, as stated in Chapter 1.

Another idea presented in Chapter 1 was the quantity known as *work* or energy. In British units, the expenditure of energy is expressed in foot-pounds, meaning the product of the number of pounds of force and the number of feet of distance through which that force acts in expending the energy. In electrical work, the metric system offers a more natural choice of units, the energy unit being called the *joule,* which is the name for the product of a metric force (called a *newton*) and the metric distance known as a *meter*. Chapter 1 also presented the idea that the *volt* represents the possibility or potentiality of expending energy per unit charge, i.e., 1 volt equals 1 joule per coulomb.

112 Essentials of Electricity

For many devices the size of the *time rate of doing work* is of the utmost importance. The technical name for this time rate of doing work is *power,* and the rate is set up in appropriate units depending upon whether the British or metric system is involved. If an aircraft engine does not have enough power, it will not be able to lift the plane off the ground within the length of any reasonably long runway, or maybe not at all. In a resistor, all of the electric power goes into producing heat in the unit. Unless the physical size and shape of the resistor permit the heat to pass along fast enough to the space nearby, the temperature of the resistor will increase until the unit destroys itself. Consequently the most detailed calculations about mechanical forces and velocities or electric currents and voltages will not give useful information about practical devices unless the *power-handling capabilities* are computed also.

2. Units of Power. Power expressed in British units might be measured as a time rate for which the basic unit is the second. Then the unit of power would be foot-pounds per second. Since this unit is rather small, and since British units in use today were developed when horses were an important economic factor, power is most often expressed in British units as *horsepower,* a technical unit of such size that one horsepower equals a rate of doing work of 550 foot-pounds per second:

$$1 \text{ horsepower} = 550 \frac{\text{foot-pounds}}{\text{second}} = 33{,}000 \frac{\text{foot-pounds}}{\text{minute}}$$

In metric units, a basic quantity for measuring power is the *watt,* a name applied to the time rate of doing work specified as joules per second:

$$1 \text{ watt} = 1 \frac{\text{joule}}{\text{second}}$$

In mechanical systems the energy in joules can be expressed in terms of newtons and meters to produce relationships between forces, distances, and power. In electric circuits, the desired relationship is between power in watts, and voltage and current. It is possible to rearrange the units in the definition of the watt to read:

$$1 \text{ watt} = 1 \frac{\text{joule}}{\text{second}} = 1 \frac{\text{joule}}{\text{coulomb}} \times 1 \frac{\text{coulomb}}{\text{second}}$$

If the words "coulomb" are treated like numbers in a set of fractions, they will cancel out when the fractions are multiplied together, thereby showing that the basic definition has not been altered by their insertion.

7: Electric Power and Energy

However, the ratios expressed by these last two fractions were defined in Chapter 1:

$$1 \text{ volt} = 1\frac{\text{joule}}{\text{coulomb}}, \quad \text{and} \quad 1 \text{ ampere} = 1\frac{\text{coulomb}}{\text{second}}$$

Therefore, it is only necessary to multiply together the values of voltage and current for a resistor to get the value of the power expressed in watts for that resistor.

Example 1. What is the power required by a 110-volt 0.5-ampere incandescent lamp?

Since
$$\text{watts} = \text{amperes} \times \text{volts},$$
then
$$\text{power} = 0.5 \times 110 = 55 \text{ watts}$$

Example 2. What is the power consumed by a motor running on a 220-volt circuit, if the motor requires 4 amperes?

$$\text{watts} = \text{amperes} \times \text{volts}$$
$$= 4 \times 220$$
$$\text{power} = 880 \text{ watts}$$

Custom and tradition have set up both the British and metric systems of units in the United States. Power for automobiles and other mechanical devices is usually expressed in horsepower, whereas most electric devices have their power capabilities expressed in watts. Consequently, it sometimes appears that horsepower cannot be used to describe electric devices and that the mechanical power of automobiles cannot be described in watts. Such an idea is wrong, for both units merely express the time rate of doing work in different systems of specification. Certainly these differing systems create confusion, especially because the British system introduces confusing multipliers in most of its relationships. Today the countries of Europe, practically all of South America, and the large complex of the U.S.S.R. use the metric system. The United States, England, and countries formerly colonies of England retain the British units in commerce, while carrying on scientific work in metric units. In view of this status of units in America, it should behoove everyone to learn the metric system, even if everyone has to learn two systems of units for the present. The necessary conversions between British and metric units of power will be discussed later in this chapter.

Prob. 1-7. A lamp bulb requires 0.218 ampere for operation at 110 volts. What power does it consume?

114   Essentials of Electricity

Prob. 2-7. An arc lamp for theatrical use has a current of 4 amperes when connected to a 110-volt supply. What power must the supply furnish to the lamp?

Prob. 3-7. How much power is taken by seven lamp bulbs if each is rated as a 60-watt lamp?

3. Three Forms of the Power Equation. In Chapter 2, Ohm's law was presented in three forms, with each form having special advantage for finding one of the three quantities in the relationship. These three forms are presented here for reference:

To find voltage,

(1) $\text{volts} = \text{ohms} \times \text{amperes}$

To find current,

(2) $\text{amperes} = \dfrac{\text{volts}}{\text{ohms}}$

To find resistance,

(3) $\text{ohms} = \dfrac{\text{volts}}{\text{amperes}}$

Because the power equation in its first form is also based on the product of two quantities yielding the third, there are similarly *three forms* for the power equation:

To find power,

(1) $\text{watts} = \text{volts} \times \text{amperes}$

To find voltage,

(2) $\text{volts} = \dfrac{\text{watts}}{\text{amperes}}$

To find current,

(3) $\text{amperes} = \dfrac{\text{watts}}{\text{volts}}$

Example 3. At what voltage must a 40-watt lamp be operated if it consumes its nominal power when its current is 0.357 ampere?

From the second form of the power equation,

$$\text{volts} = \dfrac{\text{watts}}{\text{amperes}}$$

$$\text{voltage} = \dfrac{40}{0.357} = 112 \text{ volts}$$

Example 4. What current will a 25-watt lamp require when operated on a 110-volt line?

From:

$$\text{amperes} = \frac{\text{watts}}{\text{volts}}$$

$$\text{current} = \frac{25}{110} = 0.227 \text{ ampere}$$

With Ohm's law the appropriate form must be chosen when a circuit problem is to be solved, the choice depending upon the quantity that is not known. In the same way, the appropriate form of the power equation must be chosen for the solution of problems involving power calculations.

Prob. 4-7. What is the current in a motor that consumes 17,000 watts when running on a 110-volt line?

Prob. 5-7. A motor is rated for 5 amperes when it consumes 1500 watts. For what voltage is it built?

Prob. 6-7. How many watts are consumed by an electric iron that requires 6.3 amperes from a 110-volt circuit?

Prob. 7-7. What is the current required by an arc welder rated at 1056 watts for use on a 110-volt line?

Prob. 8-7. If the motor of Prob. 5-7 required 10 amperes, but consumed the same power, for what voltage would it be constructed?

Prob. 9-7. How much power is consumed by the group of lamps in Fig. 79 if each lamp requires 0.5 ampere?

Prob. 10-7. If Motor $M1$ in Prob. 12-6 consumes 2550 watts for the same current specified in that problem, what is the voltage across the motor?

Prob. 11-7. A subway car operating where the voltage of the third rail is 550 volts consumes 21,000 watts. What current is required by the car?

Prob. 12-7. A miniature 12-volt battery constructed for a satellite can supply 45 milliamperes during the flight. What power does the battery supply at this discharge current?

4. Measurement of Power in an Electric Circuit. Because the value of power consumed in some part of an electric circuit is given by the first form of the power equation,

$$\text{watts} = \text{volts} \times \text{amperes}$$

the connection of instruments known as the *voltmeter-ammeter method* for measuring resistance will give the correct data for the power calculation. As in the measurement of resistance, care must be taken that the voltmeter indicates the operating voltage *across* just the part of the circuit where the power is to be found. Similarly, the ammeter must be *inserted into* the circuit in such a way that it will indicate only the current in the desired part of the circuit.

Since energy in an electric circuit is said to be *conserved* (energy is neither created nor destroyed when it is conserved), it must be possible

116 Essentials of Electricity

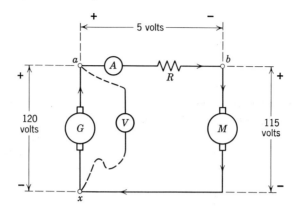

Fig. 111. Motor M is in series with resistor R.

to make a bookkeeping balance for the energy, equating the energy produced in electric form to the total energy taken away in heat or other forms. Furthermore, since power is the time rate of doing work, both input and consumed energy will be divided by the same amount of time, and therefore the bookkeeping balance can be made in terms of power in the circuit, rather than in terms of energy. This bookkeeping process is known as the *power balance* for the circuit.

Example 5. In Fig. 111, generator G sets up a current of 4 amperes around the series circuit shown. Ammeter A, inserted between the generator and series resistor R, will indicate the correct current. Connecting a voltmeter across ax, with its + terminal at a, will provide a reading for the generator potential of 120 volts. When the voltmeter is connected across ab, with its + terminal again at a, the reading is the drop in R, or 5 volts, with the drop in ammeter A assumed to be negligible. The drop across the motor is found from the reading of the voltmeter when it is connected between points b and x, with the + terminal at b. As shown, the motor drop is 115 volts from b to x. How much power does resistor R consume and what is the power taken by the motor?

Apply the power equation to each *part* of the circuit to find the power for that part. For resistor R,

$$\text{watts} = \text{amperes} \times \text{volts}$$
$$\text{power} = 4 \times 5 = 20 \text{ watts}$$

For motor M,

$$\text{power} = 4 \times 115 = 460 \text{ watts}$$

Generator G supplies the total power of $460 + 20 = 480$ watts. The power balance for this circuit may be checked by finding the power supplied by the generator:

$$\text{watts} = \text{amperes} \times \text{volts}$$
$$\text{power} = 4 \times 120 = 480 \text{ watts}$$

7: Electric Power and Energy

For many devices and circuits, an understanding of the power balance and its relationship to the various power-handling capacities of the component parts is often as important as the voltage and current calculations themselves. Often, however, the circuit calculations must be made to assist in the process of finding the power balance. When the voltage and current for each part of a circuit are not given, they must first be calculated with the aid of Ohm's law and the two Kirchhoff's laws. Then the power equation can be applied to that part of the circuit to determine the power consumed or supplied as a portion of the power balance.

Example 6. What power is consumed by a lamp bulb having 200 ohms of resistance when operated on a 110-volt circuit?

First step. (Calculate the voltage and current.) From Ohm's law:

$$\text{amperes} = \frac{\text{volts}}{\text{ohms}}$$

$$\text{current} = \frac{110}{200} = 0.55 \text{ ampere}$$

Second Step. (Compute the power.) From the power equation:

$$\text{watts} = \text{volts} \times \text{amperes}$$
$$\text{power} = 110 \times 0.55 = 60.5 \text{ watts}$$

Prob. 13-7. Generator G in Fig. 112 furnishes 1.8 amperes to the line at a potential of 125 volts. What power is consumed (a) by lamp L, (b) by the 10-ohm resistance R?

Prob. 14-7. What power does a subway-car heater use if its resistance is 220 ohms and the voltage across it is 550 volts?

Prob. 15-7. What power does an electric iron use on a 115-volt circuit if its resistance is 25 ohms?

Prob. 16-7. A 110-volt projection arc lamp requires 9.6 amperes when operating properly. What power does the lamp consume?

Prob. 17-7. What power does the motor require in Prob. 18-6?

Prob. 18-7. What power is consumed by each lamp in the group of lamps in Prob. 18-6?

Prob. 19-7. The heating element in a small electric coffee pot has a resistance of 50 ohms, and it requires 2.3 amperes. What power does the pot consume?

Prob. 20-7. The resistance of a single wire of a telegraph line is 2500 ohms.

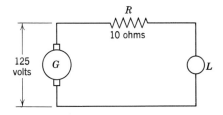

Fig. 112. Lamp and resistor in series with generator.

118   Essentials of Electricity

How much power is consumed in this wire when the steady current for the telegraph is 4 milliamperes?

Prob. 21-7. What power does the starting motor take from the battery in Prob. 6-2?

Prob. 22-7. What is the heating power needed by the vacuum tube in Prob. 8-2?

Prob. 23-7. To what heating power is the sufferer subjected for the electric-shock conditions stated in Prob. 36-2?

Prob. 24-7. For the circuit of Prob. 20-3, show the power balance by computing the power supplied by the generator and by totaling the powers required by each lamp.

Prob. 25-7. Find the power taken by the circuit of Prob. 24-3.

Prob. 26-7. How much heating power is absorbed by the copper bus bar of Prob. 42-3?

Prob. 27-7. Show the power balance for the series circuit of Prob. 44-3.

Prob. 28-7. Find the power supplied by the generator in Prob. 11-6.

Prob. 29-7. In Prob. 13-6, what is the power consumed by motor $M2$?

5. The Wattmeter. Power can also be measured by a single instrument called a *wattmeter*. A schematic drawing of the circuit of a wattmeter is shown in Fig. 113. A movable coil B is connected in series with a large resistance R, chosen so that this portion of the wattmeter functions much as the movable coil of a D'Arsonval voltmeter would act. However, instead of a permanent magnet for providing a fixed magnetic flux against which movable coil B reacts, fixed coils A (placed in series with the load current) provide an amount of magnetic flux proportional to the size of the load current. Therefore, the pointer and movable coil B experience a turning force proportional to the *product* of the current and voltage supplied to the load. With suitable calibration, the

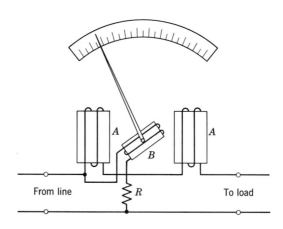

Fig. 113. Diagram of wattmeter.

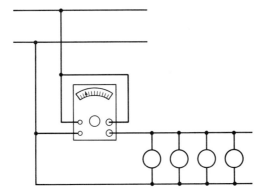

Fig. 114. Wattmeter connected to measure power taken by lamp load.

wattmeter then performs internally the multiplication specified by the power equation:

$$\text{watts} = \text{volts} \times \text{amperes}$$

Figure 114 shows how a wattmeter is connected to measure the power taken by a group of lamps. The large circles represent the terminals of the fixed current coils of the instrument. Since both the fixed-coil current circuit and the moving-coil potential circuit can be reversed, the manufacturer's instructions for polarity should be consulted and followed carefully to insure an upscale reading.

Example 7. Each of the four lamps in Fig. 114 requires 1 ampere at 112 volts. What does the wattmeter read?

The current in the current coils is $4 \times 1 = 4$ amperes, and the voltage across the voltage coil and its resistor R is 112 volts. The power consumed by the lamp load is the reading of the wattmeter, or $112 \times 4 = 448$ watts.

Prob. 30-7. Copy the diagram of Fig. 113 and label the coil terminals with dots and large circles to show the terminals that correspond to those shown in Fig. 114.

Prob. 31-7. Assume that the same wattmeter is shown in both Fig. 113 and Fig. 114 and that its potential-coil current for Example 7 is 20 milliamperes. Copy Figs. 113 and 114 and mark clearly all the sections of wire in each figure that are carrying 4 amperes.

Prob. 32-7. Show how the wattmeter of Fig. 113 should be connected to indicate the power taken by the 40-ohm load of Fig. 99.

6. Line Loss. When the resistance of the connecting line wires is important in a power-distribution circuit, an accounting of the power consumed in these wires must be made. Since the power consumed in these wires is the same as power consumed in any device, the power equation

120　Essentials of Electricity

is also applied to this special calculation. Although such power is necessary for establishing the line current through the line wires, the power consumed by the line wires is all lost as heat to the space around the wires. Therefore, since this wasted power is not available for use by the loads on the system, it is called *line loss*.

For a system having two line wires, it was stated in Chapter 6 that the *line drop* is the total potential drop for a section of line, found by adding together the voltage drops along each of the two wires for that line section. Similarly, the line loss is the power wasted in heat in *both* of the line wires of a section of the line supplying a load. This power may be calculated by finding the power lost in each wire and multiplying the answer by 2 when the resistances are the same; alternatively, the line drop (which contains the doubling factor) may be multiplied by the line current to obtain the line loss.

Example 8. Find the line loss in Fig. 115: (*a*) from the line drop, and (*b*) from the power balance.

Inspection of Fig. 115 shows that the line drop is:

$$115 - 110 = 5 \text{ volts}$$

Therefore, the line loss is:

$$\text{watts} = \text{volts} \times \text{amperes}$$
$$\text{power} = 5 \times 20 = 100 \text{ watts, answer for } (a)$$

Power balance:

generator output power:

$$\text{watts} = \text{volts} \times \text{amperes}$$
$$\text{power} = 115 \times 20 = 2300 \text{ watts}$$

motor input power:

$$\text{power} = 110 \times 20 = 2200 \text{ watts}$$

Since the motor power plus the line loss is the total power consumed and since this total must equal the generator output power, the line loss must be the difference between the generator output power and the motor input power:

$$\text{line loss} = 2300 - 2200 = 100 \text{ watts, answer for } (b)$$

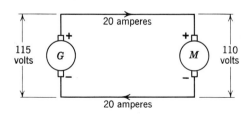

Fig. 115. The line loss is 100 watts.

7: Electric Power and Energy 121

Prob. 33-7. How many watts are consumed in each of the line wires of Prob. 29-3 if the resistance of each wire is 1.5 ohms? What is the line loss?

Prob. 34-7. What is the line loss in Prob. 18-6 for the section of line between the generator and the motor?

Prob. 35-7. What is the line loss in Prob. 20-6 for the line section between the two groups of lamps?

Prob. 36-7. Find the power lost in each of the series arms of the ladder network of Prob. 22-6.

7. Kilowatt and Horsepower. For measuring the size of blocks of power normally encountered in most household heating appliances and in practically all industrial motors, the watt is too small. Since the watt is a metric unit, it is converted into a larger unit by multiplying by a factor of 1000. In the metric system, when a unit is multiplied by 1000, the new larger unit is prefixed with the name Kilo. Long distances are measured in kilometers, large resistances are measured in Kilohms, and large powers are measured in *kilowatts*. Although the capital letter in the spelling was part of the original organization of the metric system, the examples given illustrate how common usage has modified the basic form.

Since for electric-power measurements the kilowatt has become as important a unit as the watt, their relation is important. In numbers, it is:

$$1 \text{ kilowatt} = 1000 \text{ watts}$$

Specifically:

$$2500 \text{ watts} = \tfrac{2500}{1000} = 2.5 \text{ kilowatts}$$
$$450 \text{ watts} = \tfrac{450}{1000} = 0.45 \text{ kilowatt}$$

Example 9. What power is consumed by a motor that requires 20 amperes at 220 volts?

$$\text{watts} = \text{volts} \times \text{amperes}$$
$$\text{power} = 220 \times 20 = 4400 \text{ watts}$$
$$= \tfrac{4400}{1000} = 4.4 \text{ kilowatts}$$

As noted in Section 7-2, in the United States the power capacity of mechanical devices is usually specified in horsepower. Since electric machines operate on electric power at one end and handle mechanical power at the other end, it is important to know the conversions between horsepower and kilowatts when working on machinery. Expressed in decimal notation, these relationships are:

$$1 \text{ kilowatt} = 1.34 \text{ horsepower}$$
$$1 \text{ horsepower} = 0.746 \text{ kilowatt} = 746 \text{ watts}$$

122 Essentials of Electricity

For estimating purposes, these decimals are sometimes approximated to give simple fractions in the conversions:

$$1 \text{ kilowatt} = 1\tfrac{1}{3} \text{ horsepower}$$
$$1 \text{ horsepower} = \tfrac{3}{4} \text{ kilowatt}$$

It is clear that the kilowatt represents a larger block of power than does the horsepower. Therefore, in the description of the same quantity of power, there will be more units of horsepower than the number of kilowatts needed.

Example 10. What horsepower does the motor of Example 9 consume?

$$1 \text{ kilowatt} = \tfrac{4}{3} \text{ horsepower}$$
$$4.4 \text{ kilowatts} = 4.4 \times \tfrac{4}{3} \text{ horsepower}$$
$$= 5.87 \text{ horsepower}$$

Rotating machinery is usually rated for power-handling capacity in terms of the output possible. Generators are usually rated in kilowatts for their electric output, and motors are most often rated in horsepower for their mechanical output.

Example 11. What is the output power of a 10-horsepower motor expressed in kilowatts?

$$1 \text{ horsepower} = \tfrac{3}{4} \text{ kilowatt}$$
$$10 \text{ horsepower} = 10 \times \tfrac{3}{4} \text{ kilowatts}$$
$$= 7.5 \text{ kilowatts}$$

Prob. 37-7. Express the power in kilowatts and in horsepower consumed by a 115-volt lamp that requires a current of 0.76 ampere.
Prob. 38-7. Express the answers to Prob. 13-7 in kilowatts and horsepower.
Prob. 39-7. Express the answer to Prob. 14-7 in kilowatts and horsepower.
Prob. 40-7. Express the power in Prob. 16-7 in kilowatts.
Prob. 41-7. Express the power in Prob. 19-7 in kilowatts.

8. Efficiency of Electric Devices. No machine, and in particular no electric rotating machine, produces as output power in one form the same amount of power it received in some other form. The percentage of input power that appears as output power is called the *efficiency* of the machine. Thus a motor that produces 9 kilowatts of mechanical output power for 10 kilowatts of electric input power is said to have an efficiency of 90%. If the motor produces only 8 kilowatts of output power for every 10 kilowatts of input power, it has an efficiency of only 80%.

The power supplied to a machine is often designated simply as the *input,* and the power produced by the machine likewise is simply called the *output.* In these terms, the efficiency may be written as:

$$\text{efficiency} = \frac{\text{output}}{\text{input}}$$

7: Electric Power and Energy 123

Since the output is always smaller than the input, the fraction (output/input) is always less than unity, and it is stated in percentages, as 75%, 90%, etc. Another way of stating the same fact is: The efficiency of any device is always less than 100%. Of course, the output and the input must always be stated in the same units. There is no point in comparing the input of a motor in kilowatts to the output in horsepower. Each quantity must be reduced either to kilowatts or to horsepower.

Large motors and generators permit design features that help to raise the efficiency to high values near 90%. Small machines usually have restrictions on weight and size that have been balanced against the amount of wasted power they require. Consequently, many small motors have efficiencies that lie near 50%.

Example 12. A 5-horsepower motor requires 4.8 kilowatts of input to produce its rated output. What is the efficiency of the motor?

Reduce both input and output to horsepower:

$$4.8 \text{ kilowatts} = 4.8 \times (\tfrac{4}{3}) = 6.4 \text{ horsepower}$$

$$\text{efficiency} = \frac{\text{output}}{\text{input}}$$

$$= \frac{5}{6.4} = 78.2\%$$

Otherwise, reduce both input and output to kilowatts:

$$5 \text{ horsepower} = (\tfrac{3}{4}) \times 5 = 3.75 \text{ kilowatts}$$

$$\text{efficiency} = \frac{\text{output}}{\text{input}}$$

$$= \frac{3.75}{4.8} = 78.2\%$$

Prob. 42-7. A motor having an efficiency of 85% requires an input of 2 kilowatts at a certain load. What horsepower does it deliver?

Prob. 43-7. What power in kilowatts is required to operate a 12-horsepower motor having an efficiency of 90%?

Prob. 44-7. What current will the motor of Prob. 42-7 require if the motor is built for 115 volts?

Prob. 45-7. For what voltage is the motor of Prob. 43-7 intended if it requires 45.2 amperes to produce its rated horsepower?

Prob. 46-7. When driving a load, a 110-volt motor has an efficiency of 85% and requires 25 amperes. What horsepower is supplied to the load?

Prob. 47-7. What full-load current does a 1-horsepower 115-volt motor require if its efficiency is 70%?

Prob. 48-7. A diesel engine supplies 178 horsepower to a generator when it is delivering 210 amperes to a line operating at 550 volts. What is the efficiency of the generator?

Prob. 49-7. A generator delivers a current of 75 amperes to line wires across which the potential is 110 volts. What power does the generator supply, expressed (a) in kilowatts, and (b) in horsepower?

Prob. 50-7. Commercial a-c power feeds a synchronous motor, which supplies 150 horsepower to a generator while it is delivering 173 amperes into bus bars that are maintained at a potential of 550 volts. What is the efficiency of the generator?

9. Work and Energy. Kilowatthour. When power is purchased for domestic or commercial use, a basic charge is made for the *energy* consumed, or the work actually accomplished. Therefore, this cost does not only depend on the power, i.e., the rate of doing work, but rather on the product of the rate and the time during which the rate was working. Of course, this product is work, measured in foot-pounds or joules, as stated in Chapter 1. In purchasing energy, however, other units are used. For electric energy, probably the principal form of energy purchased now, the basic energy unit is the *watthour*. This quantity of work is the energy expended when work is done at the rate of 1 watt over a period of 1 hour.

Just as the kilowatt is a more practical unit of measure for power than the basic unit of the watt, the electric energy consumed both at home and in industry is usually measured in *kilowatthours*, i.e., 1 kilowatt acting over a period of 1 hour. Billing for energy costs includes a basic charge for the kilowatthours consumed, at prices usually lying in the range between 1 cent and 6 cents per kilowatthour.

Example 13. A generator delivers 2 kilowatts to a consumer for 40 hours. How much electric energy has the customer received?

$$\text{kilowatthours} = \text{kilowatts} \times \text{hours}$$
$$\text{energy} = 2 \times 40 = 80 \text{ kilowatthours}$$

Prob. 51-7. How much electric energy is consumed in 40 hours by a motor having an input of 18 kilowatts?

Prob. 52-7. What will be the cost of using 48 kilowatts for 30 hours at 5 cents per kilowatthour?

Prob. 53-7. For how many hours can a 200-kilowatt motor be run on $50.00 if electric energy costs 2 cents per kilowatthour?

10. The Watthour Meter. Energy use can be calculated if the power remains constant for reasonable periods of time. The power is multiplied by the length of time the power remains constant, and then these portions of energy are added to give the total consumption. Whereas the idea is straightforward, providing bookkeeping machines for recording the necessary information and processing the data would seem to involve computer technology and expense that could not be justified.

Instead, meters are provided to totalize the energy consumption on a continuous basis. Known as *watthour meters*, these devices register the product of kilowatts and hours to give a reading in *kilowatthours*, even though the power consumed varies between wide limits over short periods of time. Readings of these meters are usually obtained monthly, and the energy costs are then billed after computation of the costs from the difference of the monthly indications.

Although most energy today is purchased in a-c form, d-c watthour meters are found for some applications. The watthour meter is essentially a small electric motor with a permanent-magnet brake, and it has a shaft that turns at a speed proportional to the kilowatts of power passing through the device. In one commercially-available form, the forces of motor action are produced in a disc that rotates inside a molded plastic housing that is filled with mercury. The mercury helps to float the disc (of light, highly conducting material), so that very little pressure occurs on the bearing for the disc. Also, the mercury provides a low-friction sliding contact to carry the metered current into the disc, where the motor action occurs. A simplified sketch of such a device is shown in Fig. 116. Coils A are connected to the potential to be multiplied with the metered current to give the energy indication over a period of time.

Since there is no restraining spring on the movable member of this

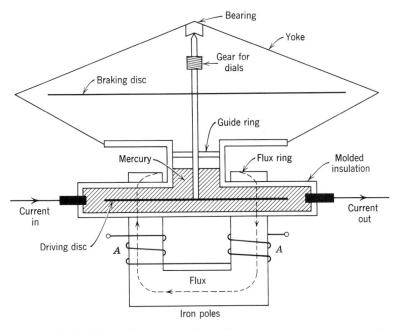

Fig. 116. Sketch of internal mechanism for a direct-current mercury-motor watthour meter.

watthour meter, the shaft turns continuously in one direction as long as the load connected through the watthour meter is consuming power. The disc in the mercury acts as a motor to drive the shaft, which in turn is geared to drive a set of recording dials.

If the shaft of the watthour meter were completely unrestrained, the speed of the shaft would keep increasing even for low power. In order that the dials keep a faithful record, they must increase their reading at a constant speed when the power passing through the meter is constant. This requirement means that there must be a braking force on the shaft

Fig. 117. Direct-current mercury-motor watthour meter. *Sangamo Electric Company.*

of the meter such that the braking force increases with speed. Then at any given power, equilibrium is achieved, at a constant speed, between the turning force on the shaft and the braking force.

A permanent-magnet eddy-current brake can produce this kind of retarding force. In Fig. 117, showing a commercial d-c watthour meter, an aluminum brake disc can be seen just above the recording dials. In the back of the case are permanent magnets that maintain magnetic flux lines through the disc. When the aluminum disc is at rest, no force is exerted upon it, even though magnetic flux from the permanent magnets passes through the disc. As the disc turns, motor action is exerted upon the natural electric charges in the aluminum, and the charges move. The currents that result are called *eddy currents* because of their whirlpool-like paths in the aluminum, and they produce a small amount of heat in the resistance of the disc. Therefore, energy is required to produce this heat, and the energy comes from a retarding force developed in reaction on the shaft of the meter. As the speed increases, the currents and the retarding force also increase. This increase in braking effort with speed is the type of retarding force needed to make the watthour meter indicate energy correctly.

At light loads, friction may cause the meter to stick so that no energy is recorded. Another circuit, not shown in Fig. 116, sends through the meter disc immersed in the mercury a small correction current in parallel with the main load current. This correction current develops a small turning force to overcome friction. As in any good compromise this balancing effect must not be made too strong. If the correction element is not properly adjusted, this antifriction circuit may cause the meter to creep slowly even when there is no current in the load circuit.

SUMMARY

POWER is the time rate of doing work; in metric units, power is measured in WATTS.

For electric quantities, power in WATTS EQUALS VOLTS × AMPERES.

The KILOWATT is a larger unit of power: 1 kilowatt equals 1000 watts.

Because mechanical power is often specified in horsepower, the important conversion is: 1 HORSEPOWER EQUALS 0.746 KILOWATT.

When line drop occurs in a distribution system, the power that is wasted in heat is called LINE LOSS.

MEASURING power with an ammeter and voltmeter or with a wattmeter requires the same care in connections as does the voltmeter-ammeter method of measuring resistance.

EFFICIENCY is the fraction of the power put into a device which is delivered by the device. This fraction is usually expressed as a percentage. Since a machine never delivers all the power put into it, the efficiency is always less than 100%.

128   Essentials of Electricity

The basic metric unit of energy is the joule; ELECTRIC ENERGY is measured commercially in kilowatthours; energy in kilowatthours equals (power in kilowatts) × (time in hours).

PROBLEMS

Prob. 54-7. A 110-volt motor requires 35 amperes. How much will it cost to run this motor if energy costs 4 cents per kilowatthour?

Prob. 55-7. How much electric energy is changed into heat when a 60-watt lamp is burned for 20 hours?

Prob. 56-7. At 6 cents per kilowatthour, what is the cost of running the iron in Prob. 6-7 for 6 hours?

Prob. 57-7. For each of the series arms of the ladder network of Prob. 23-6, find the power that goes into heat.

Prob. 58-7. A certain electric lamp requires 0.232 ampere at 108 volts. What is its rating in watts?

Prob. 59-7. When a resistor is absorbing 46 watts, its current is 2 amperes. What is the value of the resistance?

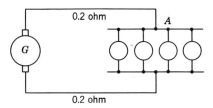

Fig. 118. Not all the power delivered by the generator is received by the lamps.

Prob. 60-7. When a 15-ohm resistor is placed across 220 volts, what power does it consume?

Prob. 61-7. How many watts does a 15-ohm resistor consume when its current is 21 amperes?

Prob. 62-7. What work is done in maintaining for 12 hours a current of 100 amperes in a wire with a resistance of 1.6 ohms?

Prob. 63-7. Lamp bank A in Fig. 118 requires 3 amperes at 110 volts.
(a) What power is lost in the line?
(b) What power is consumed by each lamp?
(c) What power is delivered by the generator?

Prob. 64-7. The efficiency of the generator in Prob. 63-7 is 95%. What is its driving power, expressed in horsepower?

Prob. 65-7. Electric energy for lighting costs 6 cents per kilowatthour. What is the cost per month (30 days) of operating a 60-watt bulb for 3 hours per day?

Prob. 66-7. Electric energy is supplied at 4 cents per kilowatthour for driving a 25-horsepower motor, having a full-load efficiency of 83%. Find the cost of operating the motor for 100 hours.

Prob. 67-7. A bill for electric energy was $16.40 for 360 hours. If the price was 3 cents per kilowatthour, what was the average power used?

Prob. 68-7. When 100 lamps had been burned on a 110-volt circuit for 8 hours, a bill of $4.00 was presented, computed at the rate of 4 cents per kilowatthour. What was the average current required by each lamp?

Prob. 69-7. What is the total energy lost in the line wires per month of 120 hours in Fig. 89?

Prob. 70-7. The resistance of each of two lamps is 250 ohms. At 6 cents per kilowatthour, how long can the two 110-volt lamps be burned for $1.00?

Prob. 71-7. What is the efficiency of a transmission line that receives 5 kilowatts at 115 volts and delivers power at 110 volts?

Prob. 72-7. How much power is delivered by the transmission line in Prob. 71-7? What is the line loss?

Prob. 73-7. The electric heater shown in Fig. 119 is intended for marine installation on a 115-volt system. The power consumption is 4 kilowatts.

(*a*) How much power is required if three such heaters are operated in parallel?

(*b*) If all three of the heaters are operated for an average of 8 hours per day, how much energy do the heaters consume in one day?

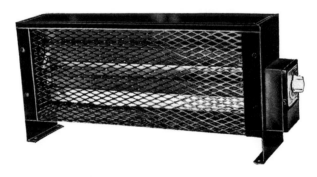

Fig. 119. Gradient-convection heater. *Acme Electric Heating Corp.*

Prob. 74-7. The lamps in Fig. 85 are burned for 3 hours a day. At an energy cost of 3 cents per kilowatthour, what is the cost per year of 300 days for the transmission losses?

Prob. 75-7. A 110-volt lamp has a resistance of 121 ohms. How much power does it consume?

Prob. 76-7. An electric clock requires 2 watts. How much does it cost to run it per year (24 hours for 365 days) if electric energy sells for 5 cents per kilowatthour?

Prob. 77-7. Operating an electric arc for a spectrograph usually requires a resistor in series with the carbon gap where the arc occurs. This resistor, which controls the arc, is called a ballast resistor. When operated from a 110-volt source, the arc itself requires 9.5 amperes at 52 volts. How much power is consumed by the arc and how much power is wasted in the ballast resistor?

Prob. 78-7. If electric energy costs 6 cents per kilowatthour, what is the cost of burning five 75-watt bulbs for 6 hours?

Prob. 79-7. How many 60-watt bulbs can be burned for 100 hours when the energy cost is 6 cents per kilowatthour, if the total bill is not to exceed $2.00?

Prob. 80-7. What is the full-load efficiency of an 8-horsepower motor that requires 63 amperes at 112 volts?

Prob. 81-7. When a generator with an efficiency of 88% is receiving 100 horsepower from its diesel-engine drive, what current is it supplying to a system operating at 220 volts?

Prob. 82-7. What is the efficiency of a 7.5-horsepower motor if its input is 7.5 kilowatts?

Prob. 83-7. The generator in Fig. 120 delivers 7 kilowatts to the line. The motor consumes 6.6 kilowatts. Find (a) the line loss, (b) the current and voltage of the motor, and (c) the terminal voltage of the generator.

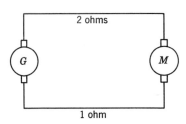

Fig. 120. Generator G must deliver enough power to supply line losses as well as power required by motor M.

Prob. 84-7. The input to a motor is 14 kilowatts, supplied over two wires having 0.08 ohm of resistance each.

(a) How much power is lost in the wires if the voltage of the motor is 110 volts?

(b) What power is lost in the wires when the voltage of the motor is 220 volts?

Prob. 85-7. What is the generator potential for each of the conditions in Prob. 84-7?

Prob. 86-7. At 3 cents per kilowatthour, what is the cost of the wasted power in Prob. 84-7, calculated per year of 300 days of 10 hours each?

Prob. 87-7. A subway car requires an average current of 60 amperes at 550 volts in order to maintain an average speed of 20 miles per hour, including stops. If energy can be provided at a price of 1.5 cents per kilowatthour, what is the energy cost of operating the subway car for one mile?

Prob. 88-7. An air conditioner is rated 7.5 amperes at 115 volts. If the price of energy is 3 cents per kilowatthour, what is the energy cost of continuous cooling for 24 hours?

Prob. 89-7. If the motor in the air conditioner of Prob. 88-7 is rated at 1 horsepower, what is the efficiency of the motor?

8

Batteries

1. Generators versus Batteries. There are three ways of commercially providing voltage sources for d-c circuits.

Chemical action permits electric charges to be separated to provide a voltage. Voltage sources employing this principle are referred to as *batteries*.

Electric machines may be constructed in which the motion of wires through magnetic fields causes charges in those wires to be separated by the forces described in Chapter 4 as *motor action*. When a commutator is provided in such machines to keep the polarity of the terminals fixed, these machines are called *d-c generators*.

However, electric power for most applications can be generated and transmitted more economically when the voltage varies with time to produce a reversing current known as alternating current. The generators of such polarity-reversing voltages are known as a-c generators, or alternators. For circuits that require d-c operation, the a-c form of energy must be converted to d-c form, in devices known as rectifiers. Most of the practical rectifiers utilize electronic principles of control over charge motions, or the closely related principles of charge motion within solid-state semiconductors. This third scheme of producing voltages for d-c circuits is called the *alternator-rectifier* method, or usually simply *rectification*.

When large quantities of d-c power are required, a cost study must be made to determine whether to drive d-c generators with a-c motors, or to employ a rectifying scheme. For large steel-mill motors, d-c power requirements are often provided from d-c generators. Electric railways, usually operating most effectively with d-c power for the driving motors, often provide the d-c power with electronic rectifiers, either placed beside the track to energize a d-c third-rail system, or mounted inside locomotives to convert a-c power taken from an overhead-wire system.

Television receivers, as well as most communications equipment, require a d-c supply for proper polarization of their electronic and solid-state components. Semiconductor devices commonly provide the rectification from a-c energy taken from the household wiring.

The importance of dry cells for flashlights and of storage batteries for starting automotive engines is clear. These applications illustrate the need for providing electric energy in some form without wired connections to a stationary source. Portable transistorized radios, space vehicles, and electric trucks for materials handling employ batteries to meet the same need. Batteries provide standby power for emergencies on systems normally supplied with commercial a-c power, such as those in hospitals and telephone exchanges.

2. Battery Terminology. Although the parts of a battery have names that depend somewhat upon its nature, i.e., whether it is a flashlight source or an automotive source, a few names are important for both types. The *container* is the outer shell, usually made of plastic, metal, or hard rubber, in which the chemicals needed for the separation of charge are held to prevent spilling or leakage. A *cell* is a basic unit comprising positive and negative *terminals* and suitable *electrodes* (usually metallic) that are immersed in a chemical solution, called the *electrolyte*. In "dry" cells, the electrolyte is mixed into a paste with other materials to assist in preventing leakage.

When several cells are connected together electrically, the combination is known as a battery of cells, or simply *battery*. For power in satellites, small wafers may be constructed to provide a basic voltage; the wafer cells are then assembled into a larger package to provide a battery able to supply both the necessary voltage and current. The battery for many flashlights is often in two separate units: two dry cells are placed end to end (in series) to provide the voltage desired for the lamp. Present-day 12-volt automobile batteries often have six cells arranged in series in a single plastic container.

Certain cells are designed to be used as long as chemical action continues efficiently, and then they are thrown away. Such cells are called *primary* cells, to be distinguished from those designed to be restored to their original chemical state by *charging,* i.e., sending a current in at the + terminal. Such rechargeable cells are called *secondary,* or *storage* cells.

3. Internal Voltage. When electrodes are immersed in a suitable electrolyte, the ensuing chemical action brings about a separation of charge. If a voltmeter having a large figure-of-merit in ohms per volt is connected between the terminals of the electrodes, the energy with which the charges are separated can be measured in volts. The voltage reading obtained is said to be the *open-circuit voltage,* since the battery cell being tested with a voltmeter supplies only a negligible current to the voltmeter. This voltage is also known as the *electromotive force,* sometimes abbreviated as emf, because the charges inside the cell must have had a force acting on them to experience the separation between the electrodes.

The emf produced by a cell depends entirely upon what electrode materials and electrolyte are used. The size of the electrodes makes no difference. A large battery cell gives exactly the same emf as a small cell made up of the same materials. Figure 121 lists a few of the combinations that are commonly considered for cells.

Since the emf of a cell is essentially a constant for the electrodes and electrolyte involved, the open-circuit voltage of the cell is also referred to as the *internal voltage* of the cell. Note that, when a cell is discharging, the current in the external circuit is directed out of the + terminal of the cell, through the circuit, and into the − terminal. Inside the cell, the charges are moving from the − electrode toward the + electrode under the influence of the internal voltage, produced by chemical action.

4. Internal Resistance. Although the size of a cell does not determine its internal voltage, the dimensions of the cell do have a close relationship to the amount of current that the cell can create when connected to a given external circuit. If the electrodes have small area, the current is smaller than if the electrodes are larger, other things being equal.

− *Plate*	*Electrolyte*	+ *Plate*	*Initial emf*
Zinc	Ammonium chloride	Carbon	1.5 volts
Lead	Sulfuric acid	Lead peroxide	2.1 volts
Iron oxide	Potassium hydroxide	Nickel	1.3 volts
Zinc	Potassium hydroxide	Mercuric oxide	1.4 volts

Fig. 121. Chemical combinations employed in commercial battery cells.

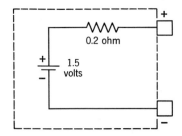

Fig. 122. The ideal-battery symbol in series with a resistance forms an equivalent circuit for an actual battery cell.

When the cell is connected into a circuit and has a current through its electrolyte, there is resistance to be accounted for in the terminals, the electrodes, and the electrolyte itself. The total influence of all these resistances can be combined into a single equivalent *internal resistance* for the cell, presumably located in series with the internal voltage. Figure 122 shows an equivalent-circuit representation for a battery cell having an internal voltage of 1.5 volts and an internal resistance of 0.2 ohm.

Once the equivalent-circuit values for a particular cell are known in the form shown in Fig. 122, calculations of its effect in setting up a current in any circuit can be made by working from the equivalent circuit. The actual cell is imagined to be removed from the circuit, and it is replaced by the series representation. Then any of the usual circuit laws may be applied as needed for the particular computation desired.

Example 1. What current will the cell of Fig. 122 set up in a 2.8-ohm resistor, having terminals marked x and y?

Figure 123 shows the cell connected to the resistance. In place of the cell, the equivalent circuit of Fig. 122 is shown connected between the terminals x and y in Fig. 124. Of course the circuit of Fig. 124 is a simple series connection, for which the solution is outlined in Chapter 3. The total resistance is $0.2 + 2.8 = 3.0$ ohms. Therefore, the current set up by the cell is 1.5 volts/3.0 ohms = 0.5 ampere.

Example 2. If the cell in Example 1 had an internal resistance of 4 ohms, what current could it establish in the 2.8-ohm resistor?

The new equivalent circuit for this cell is shown connected to the external resistor in Fig. 125. The total resistance in the circuit is $4 + 2.8 = 6.8$ ohms, and the new value of current is 1.5 volts/6.8 ohms = 0.22 ampere.

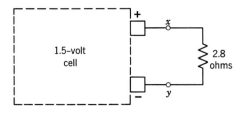

Fig. 123. An actual battery cell connected to a load resistor.

8: Batteries   135

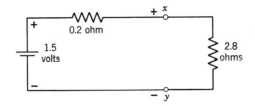

Fig. 124. The series equivalent circuit replaces the actual cell of Fig. 123.

Note how much the current was reduced when the only change in the cell was the great increase in its internal resistance.

Prob. 1-8. What current will a battery cell of 1.28 volts emf and 2.4 ohms internal resistance establish in a lamp having 12 ohms resistance?

Prob. 2-8. A dry cell has an internal potential of 1.51 volts and an internal resistance of 0.07 ohm. What current will result when a resistance of 10 ohms is connected across its terminals?

Prob. 3-8. What current will the cell of Prob. 2-8 create in a 0.1-ohm wire shunted across its terminals?

Prob. 4-8. What current will the cell of Prob. 2-8 set up in a 0.01-ohm wire placed across its terminals?

Prob. 5-8. What current will the cell of Prob. 2-8 set up in a short circuit, i.e., in a wire with practically no resistance joined to its terminals?

Prob. 6-8. A cell having the same emf as the cell of Prob. 2-8 has an internal resistance of 4.5 ohms. What current will this cell create in a 10-ohm resistor?

Prob. 7-8. What is the short-circuit current of the cell of Prob. 6-8?

Prob. 8-8. What current will be created in a 100-ohm resistor by the cell (a) of Prob. 2-8, and (b) of Prob. 6-8?

5. Terminal Voltage. In Examples 1 and 2 the terminals of the load resistors were marked x and y. These points are also the terminals of the actual cell. If a voltmeter is connected to the physical terminals of a cell, the voltmeter will read the voltage across the load resistor. This voltage, called the *terminal voltage* of the cell, is always lower than the internal emf whenever the cell supplies any current to an external circuit. Be-

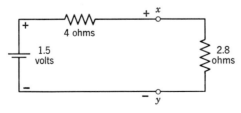

Fig. 125. Equivalent-circuit representation for a relatively high-resistance 1.5-volt cell with load.

cause the representation of the cell contains a series-connected internal resistance, the drop in this series arm of the complete equivalent circuit subtracts from the internal emf to leave a somewhat lower terminal voltage. Note that the only way to measure the internal voltage of a cell is to make sure that a voltmeter reading is taken on open circuit. Then, with zero amperes of current through the internal resistance, the terminal voltage will be the same as the internal potential of the cell.

Example 3. A cell has an internal resistance of 1.2 ohms. What internal voltage drop will occur inside the cell when it is supplying 0.5 ampere to an external circuit?

This problem requires a simple application of Ohm's law to the internal resistance of the cell: the drop across this resistance is $1.2 \times 0.5 = 0.6$ volt. If the open-circuit voltage of the cell had been measured as 2 volts, the terminal voltage would have become $2 - 0.6 = 1.4$ volts when the cell was supplying 0.5 ampere.

Example 4. Assume that the cell in Example 3 had an emf of 1.48 volts, rather than 2 volts. What would its terminal voltage be when supplying 0.5 ampere?

Since the internal drop in potential will be the same as in Example 3, i.e., 0.6 volt, the terminal voltage is: $1.48 - 0.6 = 0.88$ volt.

If the equivalent circuit of a cell is kept in mind, each of these problems is seen to reduce to that of a series-circuit calculation. The internal emf of the cell is represented by an ideal battery, and the internal resistance is shown by adding a series resistance to the ideal battery. Even if the external load resistance is not known, the terminal voltage can be found when the current supplied by a given cell is known.

Example 5. A cell having an internal resistance of 0.14 ohm and an open-circuit voltage of 1.5 volts supplies 10 amperes to an unknown load resistor. What is the terminal voltage of the cell?

The drop in the internal resistance is: $10 \times 0.14 = 1.4$ volts. If the numbers of this example are applied to the form of circuit in Fig. 124, the methods of either Chapter 3 or Chapter 6 may be employed to find the terminal voltage. By Kirchhoff's voltage law, the drop across the unknown resistor must be added to the drop in the internal resistance to equal the rise in voltage represented by the internal emf of the cell. Thus, the terminal voltage is the difference between the internal voltage and the drop in the internal resistance: $1.5 - 1.4 = 0.1$ volt.

Example 6. What is the terminal voltage of the cell in Example 5 when it is supplying only 5 amperes?

The drop in the internal resistance is now only $5 \times 0.14 = 0.7$ volt. The terminal voltage is then: $1.5 - 0.7 = 0.8$ volt.

Compare Examples 5 and 6. When the current supplied by the cell is halved, the terminal voltage increases greatly. This behavior of a cell, and therefore of its equivalent circuit, may be described thus:

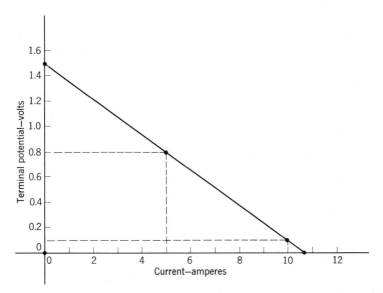

Fig. 126. A straight-line graph is called linear.

The smaller the current a cell is delivering, the greater is its terminal voltage.

This relationship can be shown clearly in a graph such as the one in Fig. 126. In the vertical direction are plotted values of terminal voltage for the cell of Example 5. In the horizontal direction are the values of current supplied by the cell. At the upper left of the graph is the point for zero current and a terminal voltage of 1.5 volts. This point represents the open-circuit voltage. The line shown in the graph slopes down to the right through the 0.8-volt terminal potential of Example 6 to the 0.1-volt terminal potential of Example 5. The short-circuit current (for zero terminal voltage) is seen to be 10.7 amperes.

Such a graph as the one in Fig. 126 is said to be linear because the path that connects all the plotted points together is a straight line. Since every cell can be represented with an equivalent circuit like that of Fig. 122, the graph of the terminal voltage of a cell will always be a straight line when calculated from this circuit. Therefore, the circuit of Fig. 122 is often called a *linear equivalent representation* for the actual cell it replaces.

Confusion over symbols sometimes arises in problems on batteries. The battery symbol can mean an actual cell, or it can mean an ideal source of voltage having zero internal resistance. In wiring diagrams of commercial apparatus, the symbol stands for actual batteries. In circuit diagrams for purposes of calculation, the symbol usually means the ideal battery. In Fig. 122, this problem was avoided by showing an outline for the cell. Throughout this chapter, a circle will be drawn around the ideal battery symbol when it means a practical cell. If there is no circle, the symbol means the ideal source of voltage having zero internal resistance. In reading commercial diagrams, however, this distinction must be kept in mind, for the two uses are seldom marked clearly.

Prob. 9-8. What is the terminal voltage of a cell which is delivering 4 amperes if it has an emf of 1.5 volts and an internal resistance of 0.2 ohm?

Prob. 10-8. Draw the linear graph representing the behavior of the cell in Prob. 9-8.

Prob. 11-8. Could the cell in Prob. 9-8 deliver 8 amperes? If so, what would its terminal voltage be? If not, why not?

Prob. 12-8. What maximum current could the cell of Prob. 9-8 supply?

Prob. 13-8. What would be the terminal voltage of the cell in Prob. 9-8 when it is delivering its maximum current?

Prob. 14-8. What would the terminal voltage be for the cell of Prob. 9-8 when it is not supplying any current?

Prob. 15-8. A cell has an internal resistance of 2.1 ohms and an emf of 1.2 volts. The external circuit has a resistance of 5 ohms. Find: (*a*) the current in the cell, (*b*) the internal voltage drop in the cell, and (*c*) the terminal voltage of the cell.

Prob. 16-8. A cell is delivering 3.2 amperes. Its internal resistance is 0.2 ohm, and its internal potential is 1.24 volts. What will a voltmeter read if put across the terminals of the cell?

Prob. 17-8. What is the external resistance in Prob. 16-8?

Prob. 18-8. Through what external resistance will a cell deliver 6 amperes if its emf is 1.4 volts and its internal resistance is 0.12 ohm?

Prob. 19-8. The cell in Fig. 127 has an emf of 1.35 volts. The ammeter reads 0.25 ampere. What is the internal resistance of the cell?

Prob. 20-8. What would a voltmeter read if put across the cell terminals in Fig. 127?

Prob. 21-8. The high-resistance voltmeter across the cell in Fig. 128(*a*) reads 1.46 volts. In Fig. 128(*b*), the same voltmeter reads 0.92 volt, and the ammeter reads 9 amperes. Find: (*a*) the emf of the cell, (*b*) the internal resistance of the cell, and (*c*) the resistance of external resistor R.

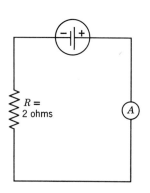

Fig. 127. Ammeter A indicates the current delivered by the battery cell.

6. Cells in Series or Parallel. When a load requires voltages or currents that are different from those that a single cell can supply, several cells may be combined into a battery of cells arranged to suit the needs

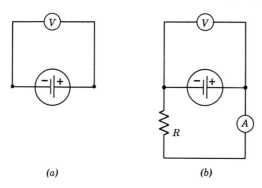

Fig. 128. Two tests determine the equivalent-circuit elements for the cell.

of the load. The simplest arrangement of a battery involves the cells in either a series or a parallel connection.

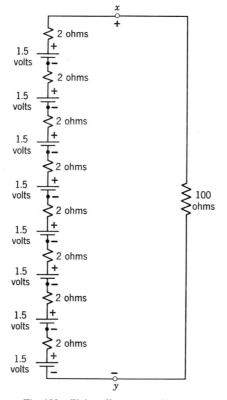

Fig. 129. Eight cells connected in series.

140   Essentials of Electricity

Example 7. Eight cells are available, each of which has an internal potential of 1.5 volts and an internal resistance of 2 ohms. What current will the series connection of these eight cells create in a 100-ohm resistor?

Each of the cells may be represented by its linear equivalent circuit, having a form like that in Fig. 122. With the actual cells replaced by their equivalents, the entire circuit connected to the 100-ohm resistor is shown in Fig. 129. Since

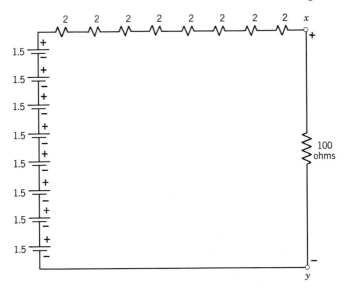

Fig. 130. All of the eight ideal internal emf's of Fig. 129 have been grouped together.

Kirchhoff's current law shows that the current is everywhere the same in a series circuit, the ideal batteries shown in Fig. 129 may all be assembled in one part of the diagram and all the internal resistances of the original cells in another, without altering the voltage or current of the 100-ohm load resistor. This rearrangement is shown in Fig. 130, where the series combinations of each kind of internal element is indicated.

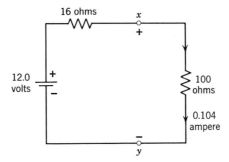

Fig. 131. The series-connected battery of eight cells has a simple series equivalent circuit.

8: Batteries   141

The battery, comprising the eight cells, has a total internal emf of 8 × 1.5 = 12.0 volts, obtained by adding together all 8 of the individual emf's. The total internal resistance of the battery is the sum of the resistances of Fig. 130: 8 × 2 = 16 ohms. Hence these total values of emf and internal resistance may be combined in a single linear equivalent representation of the battery, as shown in Fig. 131. This reduction of a battery of series-connected cells to a single linear equivalent source makes possible solving this type of problem in the same way as the problems for single cells were solved: the circuit of Fig. 131 is the same kind of simple series circuit having one source and two resistances as is the form of circuit in Fig. 124. Therefore, the total resistance of the circuit of Fig. 131 is: 16 + 100 = 116 ohms; and the current is: 12.0/116 = 0.104 ampere.

Example 8. If the eight cells of Example 7 are connected in parallel, what current will be supplied to the 100-ohm load?

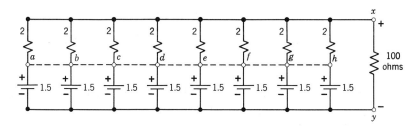

Fig. 132. Eight battery cells in parallel, each cell shown by its equivalent circuit.

The linear equivalent representations for each of the cells are shown joined in parallel in Fig. 132. A dotted line is also shown between points a, b, c, d, e, f, g, and h. Although there is no connection along this dotted line, it is drawn in to show that all the points have the same potential. If the common connection of all the negative terminals of the eight ideal battery symbols is chosen as the reference or datum point, all the points on the dotted line are positive by 1.5 volts above the datum voltage. Consequently, when each of the cells is identical, it is possible to consider that their internal emf's are in parallel even though there is no physical wire joining the emf's together.

Furthermore, since all of the internal resistances are physically joined together at point x of Fig. 132, and since the opposite end of each of the resistances is operated at exactly the same voltage along the dotted *equipotential* line, all of the internal resistances must have the same voltage across them. Therefore, the effect is that all of the internal resistances of the individual paralleled cells are in parallel, if each of the cells is identical. For the special case of a number of equal parallel resistors, Chapter 3 showed that the equivalent resistance was equal to the quotient of one of these resistances divided by the number in parallel. Thus the equivalent internal resistance of the combination of the eight cells in parallel is: $\frac{2}{8} = 0.25$ ohm.

142 Essentials of Electricity

The equivalent internal values for the battery of eight cells may be drawn as a series combination, shown in Fig. 133. Again this form is the same as that of Fig. 124. The total resistance in Fig. 133 is $0.25 + 100 = 100.25$ ohms. Thus the current supplied to the 100-ohm resistor by the eight-cell battery is: $1.5/100.25 = 0.0149$ ampere.

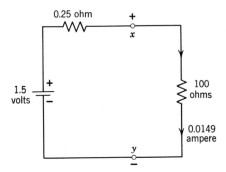

Fig. 133. The parallel-connected battery of eight cells has a simple series equivalent circuit.

7. Unlike Cells. In the two preceding Examples, it should be clear that a series grouping of cells and a parallel grouping of cells could each be reduced to an equivalent single source. The *form* of the linear equivalent representation of a single cell (with an ideal battery in series with an internal resistance) is also the *form* of the linear equivalent representation for a series or parallel grouping of identical cells. In this section the connection in series and parallel of cells that are not identical will be considered. Although such groupings of nonidentical cells are not often made practically, the behavior that does result for such connections is worth studying for the help it gives in understanding equivalent circuits.

If two or more nonidentical cells are connected together in *series*, the linear equivalent representation of the resulting battery may be found in the same ways as outlined in Example 7. The internal emf's may all be grouped together, and their sum is the emf of the linear equivalent source. Also, the separate internal resistances may be added together to find the equivalent internal resistance of the battery. However, if unlike cells are joined in *parallel*, the method outlined in Example 8 must be reconsidered.

Example 9. Two cells are to be connected in parallel to supply a 0.8-ohm load, as shown in Fig. 134. Cell A has an emf of 1.5 volts and an internal resistance of 0.25 ohm. Cell B also has an emf of 1.5 volts, but its internal resistance is 1.0 ohm. Find the current in the 0.8-ohm load.

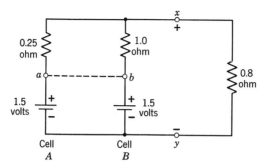

Fig. 134. Parallel connection of two 1.5-volt cells having unlike internal resistances.

Again a dotted line shows an equipotential connection between points a and b in Fig. 134. Thus, the internal emf of the linear equivalent circuit for this two-cell battery is 1.5 volts. Since the separate cell resistances are also in parallel, the equivalent internal resistance may be found by combining the separate conductances: 0.25 ohm, 1/0.25 = 4.0 mhos; 1.0 ohm, 1/1.0 = 1.0 mho. The total equivalent conductance is 4.0 + 1.0 = 5.0 mhos, leading to an equivalent parallel resistance of: 1/5.0 = 0.2 ohm. Therefore, the *form* of the linear equivalent representation of this battery with two unlike cells is still the same as that of Fig. 122, with the internal values being: emf = 1.5 volts, resistance = 0.2 ohm. Thus the current supplied to the 0.8-ohm load is found as before: the total series resistance is 0.2 + 0.8 = 1.0 ohm; and the current is 1.5/1.0 = 1.5 amperes.

Although Example 9 shows that the basic form of linear equivalent source can be applied to the parallel combination of two unlike cells, it is helpful to pursue these calculations further.

Example 10. Find the terminal voltage of the battery in Example 9, and the currents in each cell.

Observe that the terminal voltage between points x and y is also the drop across the 0.8-ohm load. Therefore, Ohm's law yields: 0.8 × 1.5 = 1.2 volts as the terminal voltage of the battery.

Kirchhoff's voltage law may then be employed to find the drop in the separate internal resistances of each cell. Choose a path from y up to a through cell A, around to x, and down through the 0.8-ohm load back to y. The drop across the 0.25-ohm internal resistance of cell A is then 1.5 − 1.2 = 0.3 volt. A similar calculation for cell B shows that the drop in its internal resistance is also 1.5 − 1.2 = 0.3 volt.

Ohm's law then gives the current in each cell. In cell A, the current is 0.3/0.25 = 1.2 amperes. The current in cell B is 0.3/1.0 = 0.3 ampere. It is instructive to note that these currents both enter point x, adding together to give the load current, or 1.5 amperes.

It is clear that the cell with the lower resistance supplies the larger part of the load current. This unequal division of the load current be-

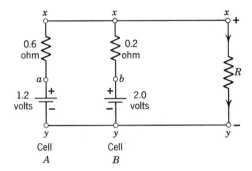

Fig. 135. Parallel connection of two cells differing in both emf and internal resistance.

tween the cells is one of the reasons for avoiding the parallel connection of unlike cells. The next example demonstrates another disadvantage.

Example 11. The two cells of the battery in Fig. 135 are not identical: cell A has an emf of 1.2 volts and an internal resistance of 0.6 ohm, and cell B has an emf of 2.0 volts and an internal resistance of 0.2 ohm. Measurement shows that the current in load resistor R is 2.0 amperes when the value of R is 0.75 ohm. Find the terminal voltage of the battery, and the current in each cell.

Ohm's law applied to the load resistor yields its drop, which must also be the terminal voltage of the battery: $2 \times 0.75 = 1.5$ volts. For the direction in which the current is observed in the load, point x is the positive terminal.

Next, apply Kirchhoff's voltage law around the following mesh: start at y, go up through cell B to x, and down through the load back to y again. In this path, the drop from point b to point x is $2.0 - 1.5 = 0.5$ volt. Thus, the current supplied by cell B is $0.5/0.2 = 2.5$ amperes. This result may seem strange, since the load receives only 2.0 amperes. However, cell B must supply 2.5 amperes to satisfy Kirchhoff's voltage law. The difference in these two values of current goes to cell A.

Note that in this problem points a and b are *not* at the same voltage. Therefore, it is not possible to find the internal emf of the battery simply by joining these points with a dotted connection as though the two cells had their emf's in parallel. Instead, the voltage law must be applied carefully around the following path: start at y, go up through the 1.2-volt rise in cell A to point a, through the internal resistance of cell A to point x, and down through the 1.5-volt drop from point x to point y. Observe that there must be a *drop* in voltage from point x to point a of $1.5 - 1.2 = 0.3$ volt. Since this drop has its *plus* sign at x and its *minus* sign at a, there must be a current in the 0.6-ohm internal resistance of cell A directed *in* at its connection to x and *out* at the cell's connection to y. Ohm's law applied to the 0.6-ohm internal resistance shows that the value of this current is $0.3/0.6 = 0.5$ ampere.

This current is being *absorbed* by cell A at the expense of cell B. In the terminology of Section 8-2, cell B is *charging* cell A at the same time that the load is being supplied from cell B. A check on the currents at point x with the help of Kirchhoff's current law verifies these results. The current entering point x is

2.5 amperes from cell B; leaving point x are currents of 0.5 ampere into cell A and of 2.0 amperes into the load.

Thus, when two cells having differing internal emf's are placed in parallel, the one with the larger emf will charge the one with the smaller emf, unless the load resistance is very low. However, if the load resistance is very small, the drop in the internal resistance of the cell having the larger emf becomes too large to be practical. Therefore, the parallel connection of cells of unequal emf is not useful.

Prob. 22-8. What is the terminal voltage of the battery in Fig. 135 when the current in cell B is 4.0 amperes? Also find the currents in the load and in cell A.
Prob. 23-8. What is the size of load resistor R in Prob. 22-8?
Prob. 24-8. Find all the currents in Fig. 135 when the load voltage is 1.6 volts.
Prob. 25-8. What is the size of load resistor R in Prob. 24-8?

8. Equivalent Circuits. Although it is not practical to connect unlike cells in parallel, a study of the *equivalent circuit* for the parallel combination of two unlike cells is useful because it illustrates the behavior of many practical devices.

Example 12. For the battery of two cells in Fig. 135, find the short-circuit current and the open-circuit voltage. Draw the graph of terminal voltage versus current supplied, including these points and the measured point in Example 11.

If the battery is shorted, resistance R is zero. Whatever current is in R will produce zero volts between the terminals x and y. With this information, Kirchhoff's voltage law gives the drops across the two internal resistances. In cell A the drop across the 0.6-ohm internal resistance is $1.2 - 0 = 1.2$ volts. For cell B, the drop across its 0.2-ohm internal resistance is $2.0 - 0 = 2.0$ volts. Ohm's law then yields the current in each cell. For cell A, $1.2/0.6 = 2.0$ amperes. For cell B, $2.0/0.2 = 10$ amperes. Therefore, since both of these currents enter point x, the current in the short-circuit connection between points x and y is the sum of the cell currents: $2 + 10 =$ **12 amperes.**

If resistor R is open-circuited, i.e., if R is completely disconnected from points x and y so that they are not connected externally to the battery, there is still a current in the cells. Trace a counterclockwise path from y up through the rise of 2.0 volts in cell B to point b, through the 0.2-ohm resistance to x, through the 0.6-ohm resistance to a, and down through the 1.2-volt emf (as a drop) to y again. For this open-circuit condition, the two internal resistances of the cells are in *series,* and the two emf's are in opposition to each other. The voltage drop from point b to point a is $2.0 - 1.2 = 0.8$ volt. Since this is the voltage across the two internal resistances in series $(0.2 + 0.6 = 0.8$ ohm), the current in the cells is $0.8/0.8 = 1.0$ ampere, by Ohm's law. Cell B is discharging, with its current of 1 ampere being used to *charge* cell A.

This result demonstrates another difficulty with parallel cells having differing emf's. Even when the battery is sitting idle, charge passes internally among the cells, wasting power in the internal resistances of the cells.

After the current has been found to be 1 ampere from point b toward point x in the 0.2-ohm internal resistance of cell B, the drop across this resistance is seen to be $1 \times 0.2 = 0.2$ volt. Point x is then positive with respect to point y by $2.0 - 0.2 = 1.8$ volts. This value is the open-circuit voltage of this battery. The answer may be checked by noting that 1 ampere directed downward through the 0.6-ohm internal resistance of cell A produces a drop of 0.6 volt. Since this drop is positive at point x, it adds to the drop in emf through cell A, yielding the drop from points x to y for this open-circuit condition as $0.6 + 1.2 = $ **1.8 volts,** checking the answer above.

In Example 11, the battery supplied a current of 2 amperes when its terminal voltage was found to be 1.5 volts. In this calculation, the terminal voltage of the battery is 1.8 volts when no external current is delivered, and 0 volts when the short-circuit current of 12 amperes is delivered. These values are plotted in Fig. 136, and a line has been drawn joining them.

Since the line drawn in Fig. 136 is a straight line, it appears that there must be a *linear equivalent circuit* of the form of Fig. 122 for this combination of two unlike cells. In fact, a mathematical proof has been devised to show that *any* combination of *any number* of resistors and ideal batteries having *any values* can ultimately be reduced to the linear equivalent form of Fig. 122 with respect to two terminals such as x and y in the Examples. This result has been stated in a form known as *Thevenin's theorem.* For this reason, the form of Fig. 122 is sometimes called the *Thevenin equivalent circuit.* In the next section the

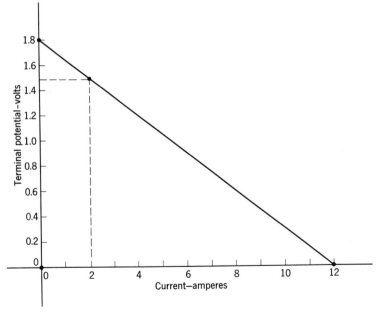

Fig. 136. Linear graph of the voltage-current relationship for the battery of Fig. 135.

method of finding the two parts of the Thevenin circuit will be considered, but a presentation of the proof of the theorem will be omitted.

Prob. 26-8. What is the short-circuit current of the cell in Prob. 1-8?
Prob. 27-8. What is the short-circuit current of the cell in Example 3 of this chapter?
Prob. 28-8. Three identical cells are connected in parallel. Each cell has an internal emf of 1.5 volts and an internal resistance of 0.2 ohm. What is the open-circuit voltage of the battery, and what is its short-circuit current?
Prob. 29-8. Two unlike cells are connected in parallel. One cell has an emf of 2.0 volts and an internal resistance of 0.1 ohm. The other cell has an emf of 1.5 volts, and its internal resistance is 0.2 ohm. What is the open-circuit voltage of this battery, and what short-circuit current will it deliver?
Prob. 30-8. The battery of Prob. 29-8 is connected in parallel with a cell having an internal resistance of 0.1 ohm and an internal potential of 1.5 volts. What will the short-circuit current of the new combination be?

9. The Thevenin Method. In any circuit containing only ideal batteries and resistances, two points of the circuit may be chosen and identified with letters, perhaps x and y. Then Thevenin's theorem states that the original circuit may always be reduced to the *form* of Fig. 137, where V_t and R_t stand for the internal emf and internal resistance, respectively, of the Thevenin equivalent circuit.

The method by which the values of the equivalent source voltage and resistance are obtained is illustrated first in an example. Application of these values to the solution of a circuit problem is demonstrated in a second example, and then the rules governing the method are stated.

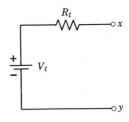

Fig. 137. The Thevenin linear equivalent circuit is a series connection of an ideal source and a resistance.

Example 13. Find the internal emf and the internal resistance of the linear equivalent circuit that represents the battery of Example 11.

For a single cell, the internal emf and the open-circuit voltage are the same. Since the values in the Thevenin linear equivalent circuit must duplicate the behavior of the battery at the chosen terminals of the battery, the internal Thevenin emf of the equivalent circuit must be the same as the open-circuit voltage of the original battery to be represented. For the battery of Example 11, the open-circuit voltage was found in Example 12 to be 1.8 volts. Therefore, the answer for V_t, the value of the Thevenin internal voltage in the equivalent circuit, is 1.8 volts.

For a single cell, the short-circuit current occurs when the external terminal voltage is zero. The value of the short-circuit current is found by dividing the internal emf of the cell by the internal resistance of the cell. Consequently, the internal resistance could be found by dividing the internal emf of the cell by its

148 Essentials of Electricity

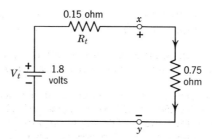

Fig. 138. The battery of Fig. 135 has been replaced by its Thevenin equivalent circuit.

short-circuit current. For the battery in Example 11, the short-circuit current was found in Example 12 to be 12 amperes. Therefore, dividing the open-circuit voltage of the original battery by the short-circuit current of the original battery will give the equivalent internal resistance R_t of the Thevenin representation. For the battery of Example 11, the equivalent internal resistance is 1.8 volts/12 amperes = 0.15 ohm. These values are shown in Fig. 138, together with an 0.75-ohm resistance, the same load resistor that was connected in Example 11.

Example 14. Find the current in the circuit shown in Fig. 138.

Since this circuit is now a simple series circuit, the solution is straightforward. The total circuit resistance is $0.15 + 0.75 = 0.9$ ohm. Then Ohm's law gives the current as $1.8/0.9 = 2.0$ amperes, checking the value originally given in Example 11.

From the steps followed in obtaining the equivalent representation of Fig. 138 for the battery of two unlike cells of Example 11, it is possible to understand the three rules governing the Thevenin method of circuit reduction. Before the rules are applied to any circuit, it is helpful to identify the two terminals between which the equivalent representation is to be made, designating the two terminals with appropriate letters or numbers.

THREE RULES FOR THE THEVENIN METHOD

First Rule. Compute the open-circuit voltage between the two specified terminals; the Thevenin voltage V_t is equal to the open-circuit voltage.

Second Rule. Calculate the short-circuit current that would be delivered at the two terminals if a wire of zero resistance were placed across them.

Third Rule. Divide the open-circuit voltage from the First Rule by the short-circuit current from the Second Rule; the Thevenin resistance R_t is equal to the quotient.

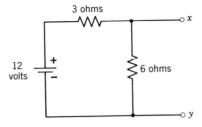

Fig. 139. This circuit may be reduced to an equivalent series circuit.

Two examples of the application of the Thevenin rules demonstrate how this method may be used to supplement the methods of circuit solution outlined in Chapter 6.

Example 15. Find the Thevenin equivalent circuit for the circuit shown in Fig. 139.

According to the First Rule, find the open-circuit voltage. Since the circuit is of simple series form, the total resistance is $3 + 6 = 9$ ohms. The current in the circuit is $\frac{12}{9} = 1.33$ amperes. The drop from x to y, the desired open-circuit voltage, is $6 \times 1.33 = 8$ volts, when rounded off. Therefore, Thevenin voltage $V_t = 8$ volts.

For the Second Rule, assume that a wire of zero resistance is connected between points x and y. Whatever current passes through it will require a drop of zero volts across it. With zero volts between points x and y, there can be no current in the 6-ohm resistor. Therefore, the 3-ohm resistor is directly across the source voltage for this calculation. The current in the 3-ohm resistor, and thus in the short circuit is $\frac{12}{3} = 4$ amperes.

The quotient specified in the Third Rule gives the internal resistance of the Thevenin equivalent circuit. Dividing the open-circuit voltage by the short-circuit current gives: $\frac{8}{4} = 2$ ohms $= R_t$. These values are shown in Fig. 140. The circuit of Fig. 140 has the same linear graph of voltage versus current for its terminals x and y as the original circuit of Fig. 139 has for its terminals x and y. Therefore, in this sense the Thevenin circuit is the linear equivalent circuit for the original circuit. The equivalence does *not* apply for power calculations.

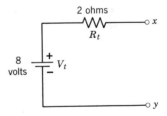

Fig. 140. The Thevenin equivalent circuit for the circuit of Fig. 139.

Example 16. With the aid of the Thevenin method, find the current in the 4-ohm load of the ladder circuit of Fig. 141.

Before the rules of the Thevenin method are applied, it is helpful to make a judicious choice of the terminals between which the Thevenin circuit is to be calculated. The terminals marked x and y in Fig. 141 represent a suitable choice for this problem. The part of the ladder circuit to the right of the ter-

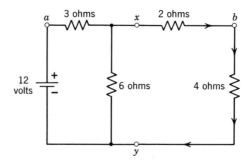

Fig. 141. A ladder circuit, with terminals x and y designated for application of the Thevenin method.

minals x and y will be temporarily set aside, leaving exactly the circuit of Fig. 139. In Example 15, the values $V_t = 8$ volts and $R_t = 2$ ohms were obtained and placed in the equivalent circuit of Fig. 140. Since this circuit behaves exactly as the original circuit would between points x and y, the right-hand part of the original ladder circuit may be rejoined to points x and y, as shown in Fig. 142. In Fig. 142, the current in the 4-ohm resistor must be exactly the same as the current in the 4-ohm resistor in Fig. 141.

However, the Thevenin method has reduced the final calculation to that for a simple series circuit. The total resistance is $2 + 2 + 4 = 8$ ohms. The current in the series circuit is 8 volts/8 ohms = 1 ampere.

The answer to Example 16 may be checked easily by the methods of Chapter 6. With 1 ampere in the 4-ohm load of Fig. 141, the voltage drop from point x to point y is 6 volts. Thus the current through the 6-ohm shunt branch of the ladder circuit is also 1 ampere. Adding the currents leaving point x shows that the current in the 3-ohm series arm is 2 amperes, producing a 6-volt drop from a to x. The summation of the two 6-volt drops around the left-hand mesh of the ladder circuit

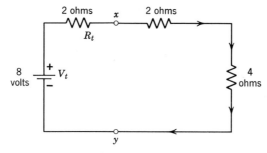

Fig. 142. Appearance of the ladder circuit of Fig. 141 after reduction with the Thevenin method.

yields 12 volts. Since the original answer of 1 ampere was calculated by the Thevenin method starting from a supply potential of 12 volts, the ladder-circuit check to show that a current of 1 ampere in the 4-ohm load requires a 12-volt supply therefore verifies the previous answer.

Prob. 31-8. Find the Thevenin equivalent of the battery of Prob. 29-8.
Prob. 32-8. Find the Thevenin equivalent of the battery of Prob. 30-8.
Prob. 33-8. Determine the Thevenin equivalent of the battery of Prob. 28-8, and plot the graph of its terminal voltage versus the current delivered by the battery.
Prob. 34-8. In the circuit of Fig. 141, choose points b and y as the terminals at which a Thevenin representation is to be found, for the purpose of determining that the current in the 4-ohm load is 1 ampere. As outlined in Example 16, remove the part of the circuit (the 4-ohm load) to the right of points b and y. Determine the Thevenin equivalent circuit for the circuit that remains connected between terminals b and y, and check that your answer will supply 1 ampere to the 4-ohm load.
Prob. 35-8. Use series-parallel network reduction to prove that the current in the 3-ohm series arm of Fig. 141 is 2 amperes.
Prob. 36-8. Prove that the current in the 2-ohm series arm of Fig. 141 is 1 ampere by removing the 2-ohm arm from the circuit, leaving a circuit between points x and b; determine the Thevenin equivalent representation of the circuit remaining between points x and b, and check that this circuit yields a current of 1 ampere to the 2-ohm arm.
Prob. 37-8. Plot the graph of terminal voltage versus current delivered for the battery of Prob. 30-8.

10. Series-Parallel Cell Groupings.

If many identical cells are grouped together in series-parallel arrangements, it is possible to arrange a battery that will have both sufficiently large emf and appropriate current-delivering capacity (i.e., suitably low internal resistance) for supplying various kinds of loads. Each of the eight cells in Example 7 had an emf of 1.5 volts with an internal resistance of 2 ohms. Suppose these are divided into two paralleled groups, each of which has four cells in series, as shown in Fig. 143, with the battery supplying a load having 4 ohms of resistance. Each of the parallel branches can be reduced to a single linear equivalent Thevenin representation, and then the two equivalent sources can be combined by means of any of the methods previously outlined.

For each parallel branch, the total emf is $4 \times 1.5 = 6.0$ volts, and the total internal resistance is $4 \times 2 = 8$ ohms, as illustrated in Fig. 144. If an equipotential connection is imagined to be drawn between the two 6-volt sources in Fig. 144, the Thevenin equivalent of the battery can be readily seen to be a 6-volt emf in series with the parallel combination of two 8-ohm resistances, or 4 ohms. When the Thevenin equivalent circuit is connected to the 4-ohm load, as shown in Fig. 145, the

152 Essentials of Electricity

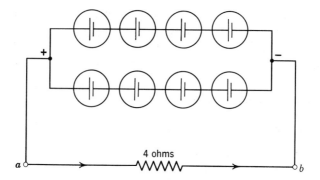

Fig. 143. Eight cells arranged in two paralleled groups of four cells in series.

total series resistance is $4 + 4 = 8$ ohms. The current in the 4-ohm load is $\frac{6}{8} = 0.75$ ampere.

If only a fixed number of cells is available, it is possible to show that the series-parallel grouping having an internal resistance for the battery nearest in size to the resistance of the load to be supplied will always provide the largest possible current for that load. However, such an arrangement may lead to considerable loss of electric power as heat inside the battery. Usually, the load requires a particular current, and cost sets a requirement on the life of a battery needed for economical operation. Therefore, *enough* cells are usually arranged in a series-parallel grouping to satisfy the latter two considerations. Nevertheless, problems based on batteries arranged from a fixed number of cells are worth studying for the practice they give in applying the various circuit methods of calculation.

Example 17. In how many ways can twelve cells be arranged?

(1) Twelve cells in series.
(2) Two parallel sets, each with six cells in series.
(3) Three parallel sets, each with four cells in series.
(4) Four parallel sets, each with three cells in series.
(5) Six parallel sets, each with two cells in series.
(6) Twelve cells in parallel.

Example 18. Find the current which each arrangement of cells in Example 17 would set up in a 1.3-ohm load resistor. Each of the twelve cells has an emf of 1.4 volts, and an internal resistance of 0.4 ohm?

Examples of the solution for the form of each of the arrangements has been given before. The values for the linear equivalent circuit for each form will be stated, with the current set up in the 1.3-ohm load. Details should be worked out to verify these answers.

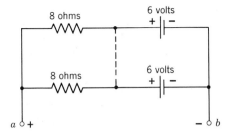

Fig. 144. Reduction of the battery of Fig. 143 to two equivalent branches in parallel.

(1) Twelve cells in series: emf = 16.8 volts; internal resistance = 4.8 ohms. Current supplied = 2.75 amperes.

(2) Two parallel sets, each with six cells in series: emf = 8.4 volts; internal resistance = 1.2 ohms. Current supplied = **3.36 amperes.**

(3) Three parallel sets, each with four cells in series: emf = 5.6 volts; internal resistance = 0.53 ohm. Current supplied = 3.06 amperes.

(4) Four parallel sets, each with three cells in series: emf = 4.2 volts; internal resistance = 0.3 ohm. Current supplied = 2.63 amperes.

(5) Six parallel sets, each with two cells in series: emf = 2.8 volts; internal resistance = 0.13 ohm. Current supplied = 1.96 amperes.

(6) Twelve cells in parallel: emf = 1.4 volts; internal resistance = 0.033 ohm. Current supplied = 1.05 amperes.

Note that combination (2), having two parallel sets, each with six cells in series, gives the largest current into the 1.3-ohm load. For this combination, the internal resistance of the Thevenin equivalent (1.2 ohms) was most nearly equal out of all the combinations to the external resistance of the load.

Prob. 38-8. What is the greatest current that twenty cells, each of 1.5 volts emf and 0.5-ohm internal resistance, can set up in a 12-ohm load? How would the cells be arranged?

Prob. 39-8. What is the greatest current that the twenty cells of Prob. 38-8 can establish in a 5-ohm load, and how would the cells be arranged?

Prob. 40-8. What is the greatest current that the twenty cells of Prob. 38-8 can set up in a 0.4-ohm resistance? How would the cells be arranged?

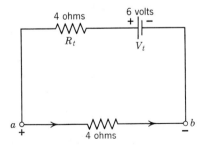

Fig. 145. The battery of Fig. 143 has been reduced to a series linear equivalent circuit.

Prob. 41-8. What is the greatest current that the twenty cells of Prob. 38-8 can set up in a 40-ohm load, and how would the cells be arranged?

Prob. 42-8. If just eight of the cells of Prob. 38-8 are available, what is the Thevenin equivalent of a battery formed with the eight cells arranged in four paralleled groups, each group having two cells in series?

11. Dry Batteries. In the familiar flashlight cell, the negative electrode is a zinc container, and the positive electrode appears as a terminal at the top center of the can. The positive electrode is a composite structure of a carbon rod embedded in a mixture of manganese dioxide and carbon particles. A moist paste of ammonium chloride and porous paper separate the two electrodes. Thus, the cell is damp, even if not wet in the sense of containing a liquid. The zinc-carbon cell is usually called a primary cell, for it is not readily rechargeable. The emf of the carbon-zinc dry cell is about 1.5 volts. As in all cells, the internal resistance depends on the size and shape of the electrodes.

The action of the electrolyte consumes the zinc by uniting with it to form a chemical compound of little commercial value. Thus the zinc may be considered a fuel, just as is the coal which is consumed under a boiler by the action of the oxygen in the air, finally formed into ash. Since the cost of the energy obtained from using zinc in a battery is much greater than the cost of the energy obtained from coal or oil burned under a boiler, the use of dry cells is limited to small-power applications where the nonspill characteristics and portability are desirable and necessary.

In practically all types of cells, the contamination of one of the electrodes with some other material creates a kind of localized cell around the impurity. When the electrolyte acts on this impurity cell, a slow internal electric discharge takes place, consuming part of the original fuel wastefully. Such a means of producing electricity is known as *local action,* a harmful property of cells that must be avoided by proper treatment of the electrodes during manufacture.

The manganese dioxide in the positive-electrode mix helps to avoid another difficulty in the operation of dry cells. When the zinc is being consumed chemically as fuel for the production of electric current from the cell, the gas called *hydrogen* is formed at the positive electrode. Since the hydrogen would soon cover the electrode and greatly increase the internal resistance of the cell, it must be removed as quickly as possible. Otherwise the terminal voltage of the cell will drop drastically because of the rapid increase of internal resistance. This collecting of hydrogen gas on the positive electrode is called *polarization.* The manganese dioxide is able to remove the effect of the hydrogen so that the cell retains a reasonably low value of internal Thevenin resistance.

8: Batteries 155

Fig. 146. This radio "B" battery has forty-five cells connected in series. *Burgess Battery Company*.

Sometimes the manganese dioxide is said to minimize the effects of polarization, or it is said to *depolarize* the cell.

Dry cells of this same chemical form can be combined into batteries in the various ways already indicated in the previous sections. Figure 146 shows a battery arranged to supply 67.5 volts for a portable radio. Other forms are readily available for the flashlights, penlights, lanterns, toys, games, and photoflash units which are virtually standard household equipment.

Mercury dry cells offer another form of packaged electricity. Figure 147 shows some of the forms of batteries and their cells of the mercury type. Electrodes of finely divided materials are pressed into appropriate

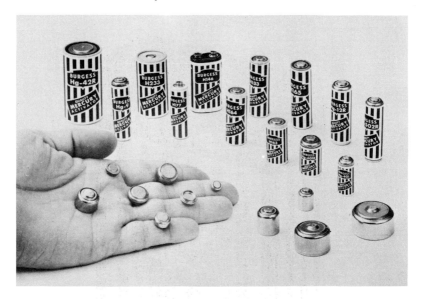

Fig. 147. Mercury cells and batteries of assorted milliampere-hour capacities. *Burgess Battery Company.*

shapes to be fitted into sealed steel cans. The positive electrode is mercuric oxide mixed with a conducting material such as graphite, and the negative electrode is powdered zinc. With a solution of potassium hydroxide held in a paste, the emf of mercury cells is about 1.40 volts. Some cells having a depolarizing ingredient suitable for use at high temperatures have a slightly lower emf. An important advantage of mercury cells over other dry types is that their terminal voltage remains quite constant during discharge until very near the end of usefulness for the cell.

12. Electrolysis and Electroplating. When a dry cell produces current, zinc is consumed as a fuel for the process through the formation of a chemical compound with the electrolyte. If, however, a similar cell is arranged so that current can be established through the cell from an external source, zinc can be taken from a suitable electrolyte and deposited upon one of the electrodes. This process is called *electrolysis.*

The amount of metal which 1 ampere will deposit from the solution in 1 hour is exactly the same as the amount that would have been consumed as fuel if the electrolytic cell had been supplying a current of 1 ampere for 1 hour. This amount is called the *electrochemical equivalent* of the metals, having specific values for each metal. A few typical values follow:

For zinc, the electrochemical equivalent is 0.0429 oz.
For copper, the electrochemical equivalent is 0.0418 oz.
For nickel, the electrochemical equivalent is 0.0386 oz.
For silver, the electrochemical equivalent is 0.1421 oz.

Example 19. A battery runs for 80 hours delivering current at an average value of 2 amperes. How many ounces of zinc are consumed?

1 ampere for 1 hour consumes 0.0429 oz.
2 amperes for 1 hour consume $2 \times 0.0429 = 0.0858$ oz.
2 amperes for 80 hours consume $80 \times 0.0858 = 6.86$ oz.

Prob. 43-8. How much zinc is consumed when a battery delivers 0.48 ampere for 145 hours?

Prob. 44-8. The zinc plate of a battery weighs 8 ounces. How long will it last if the battery delivers an average current of 0.7 ampere for 5 hours every day?

Prob. 45-8. The zinc plate of a battery weighed 20.1 ounces before any current was taken from the cell. After a run of 200 hours, the plate weighed 13.7 ounces. If no local action wasted any zinc, what average current was taken from the cell?

The process of depositing metal from solutions by passing a current through the solution is called *electroplating,* or often simply plating. Thus, as in Fig. 148, a piece of copper and a piece of iron may be immersed in a solution of copper sulfate. If generator G establishes a current through the liquid from the copper to the iron, the copper will be separated from the solution of copper sulfate and will be deposited on the iron. The iron will become copper plated. As fast as the copper goes out from the solution and is deposited upon the iron, more copper comes off the piece of copper and goes into solution to replace that which went to the iron from the solution.

For a silver plate over the iron, a piece of silver in an appropriate silver solution must be connected. A nickel solution and a nickel strip will plate the iron with nickel plate.

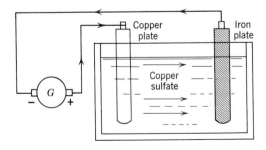

Fig. 148. The iron becomes copper plated when the generator establishes current through the electrolytic cell in the direction indicated.

158 Essentials of Electricity

13. Electrotyping. The object of electrotyping is to reproduce whole pages of previously set-up type, engravings, etc. First a wax impression is taken of the set-up type to be reproduced. Since wax does not conduct electricity, the wax mold is given a coating of graphite, or carbon in powdered form. The whole mold with its coating of graphite is immersed in copper sulfate together with a copper bar. An electric current is established from the copper bar to the graphite coating, causing copper to be deposited on the graphite, just as the iron was copper-plated in the previous section. The current is left on long enough to deposit a plate of sufficient thickness for safe handling. After the wax mold is removed, the copper plate remaining is an exact reproduction of the original set-up type. The copper plate then may be backed up with stronger metal before this electrotype is used for printing.

14. Refining of Metals. Since the metal coating produced by plating is remarkably pure, this process is often used to separate metal from impurities. The impure mass is made the positive electrode of an electrolytic cell. When a current is set up in the cell from an external source, the pure metal is gradually dissolved by the electrolyte and carried over to the negative electrode. At the end of the process the negative electrode is found to consist of very pure metal, since practically all the impurities remain at the positive electrode.

Prob. 46-8. How long will it take to refine 175 pounds of copper if a current of 100 amperes can be used?

Prob. 47-8. An iron casting is to be copper plated and then nickel plated. The current for each plating is to be 10 amperes. How long must the casting remain in each vat in order to have 14 ounces of each metal deposited on it?

Prob. 48-8. Two electroplating vats are arranged in series, one for nickel plating and the other for silver plating. If the current through the vats deposits 2 ounces of silver in a given time, how much nickel is deposited at the same time?

15. Electrolytic Destruction. Whenever metals are located in an electrolyte, the presence of any electric current can bring about destruction of the metal. The source of the current may be the *local action* of dissimilar metals in contact, or the current may come as a leakage from a circuit near the metals. The destruction of local action is constantly a problem for ships in the saline solution of the sea. Electric railways have often faced litigation as the source of destructive electrolysis of nearby water pipes or gas mains.

The track of electric railways is usually the return path for current from the traction motors back to the generators of the system. Since the rails are not usually insulated from the earth, some current will find a lower-resistance, parallel path through the earth back to the generator, rather than remaining entirely confined in the running rails for the entire return to the generator, and especially if the negative terminal of the generator is grounded. If a water pipe runs nearby, some of this leakage current will enter the pipe where the current leaves the rails, generally returning to the rails near the generator connection, or to the grounded negative connection of the generator itself. At the point where the current leaves the metal pipe, the pipe in moist earth acts as the positive electrode in an electroplating vat. Metal is actually plated away from the pipe, usually being deposited upon some other substance in the current path through the earth. Clearly this damage to water and gas pipes must be prevented.

The running rails of electric railways should be carefully bonded together to bring the resistance of the rail path to as low a value as possible. Sometimes an auxiliary generator is arranged to maintain the running rails at a small negative voltage with respect to the earth, essentially to ensure that the return current from the traction motors remains in the rails without leaking out to nearby pipes.

16. Automotive Batteries. One of the most important functions of an automotive battery is to provide electric current for operating the d-c starting motor that cranks the gasoline engine of the automobile. Energy must be *stored* in some form while the auto is not running, ready to be applied to this cranking task at any time. Therefore, batteries suitable for automotive use are *storage batteries,* made up of *secondary* or rechargeable cells to obviate the cost of throwing away a primary battery after one discharge.

Remember that a storage cell does not store electricity, but rather it stores chemical energy. In charging, electric energy is transformed into chemical energy and stored in the cell; in discharging, this chemical energy is changed back again into electric energy.

Of course there are losses in both transformations which prevent an efficiency of 100%. During charge, the terminal voltage *applied* to the cell must be *higher* than the emf of the cell in order that current will enter the cell through its internal resistance. When the cell discharges, the terminal voltage drops *lower* than the emf, as seen in the previous sections. In other words, power is lost in the internal resistance both on charge and discharge. Storage batteries thus do not return as much energy as is put in upon charge. The cost of this energy lost each charging cycle and the cost of ultimately replacing a storage battery are the principal components to be considered in estimating the actual cost of using a battery in place of some other means of storing energy.

For automotive batteries, lead and lead peroxide have been found to

160   Essentials of Electricity

work well in a storage cell with a solution of sulfuric acid as the electrolyte. Figure 149 shows the construction of a 12-volt battery. At full charge, the electrolyte is 35.6% by weight of sulfuric acid, the rest water. Because a given volume of sulfuric acid weighs more than the same volume of water, the 35.6% of acid added to pure water makes the electrolyte of the automotive battery weigh more than the same total volume of water would weigh. This increase of weight can be measured by a floating device called a *hydrometer,* which shows the relative increase in weight on a scale called *specific gravity.* Pure water is assigned a specific gravity of 1.000, and the sulfuric-acid battery electrolyte has a specific gravity of 1.260 at full charge. Measuring the specific gravity of a lead-acid battery with a hydrometer gives a good indication of the state of the charge in the battery.

For the electrode materials used, the lead-acid cell with its electrolyte at a specific gravity of 1.260 has an internal voltage of 2.10 volts. Therefore, such a storage cell is usually called a 2-volt cell. The battery of Fig. 149 has six of these cells in series to provide the nominal 12-volt terminal potential. In order to keep the internal resistance low, the electrodes have grids containing spongy pastes into which the electrolyte can penetrate easily. Positive plate 13 in Fig. 149 has lead peroxide pasted into the grid structure, while negative plate 14 has a porous mass of spongy lead in its grid structure. Low internal resistance is extremely important, for the starting current needed in winter with a 12-volt system usually lies between 225 and 500 amperes, depending upon the size of the gasoline engine and the viscosity of its lubricating oil.

Example 20. A lead cell has an emf of 2.05 volts and an internal resistance of 0.004 ohm. What will its terminal voltage be when the cell is discharging a current of 20 amperes?

The drop in the internal resistance is $0.004 \times 20 = 0.08$ volt. This drop subtracts from the internal emf to give the terminal voltage: $2.05 - 0.08 = 1.97$ volts.

Example 21. What external voltage would have to be impressed across the storage cell of Example 20 in order to *charge* it with 20 amperes?

The internal drop across the internal resistance is still $0.004 \times 20 = 0.08$ volt, but this drop now *adds* to the internal emf in accordance with Kirchhoff's voltage law. Therefore, the terminal voltage required for charging at the 20-ampere rate is $2.05 + 0.08 = 2.13$ volts for the cell.

Prob. 49-8. A lead storage cell has an emf of 2.04 volts and an internal resistance of 0.0025 ohm. The normal charging current is 35 amperes. What is the terminal voltage of the cell when charging begins?

Prob. 50-8. What is the terminal voltage of the cell in Prob. 49-8, if it is discharged from its original condition at twice the 35-ampere rate?

8: Batteries 161

Fig. 149. Cutaway view of an automotive battery. *The Electric Storage Battery Company.*
1. Terminal post
2. Vent cap
3. Sealing compound
4. Cell cover
5. Filling tube
6. Electrolyte level mark
7. Inter cell connector welded to . . .
8. Lead insert in cover and . . .
9. Plate strap
10. Separator protector
11. Negative plate
12. Separator
13. Positive plate
14. Negative plate with active material removed to show . . .
15. Plate grid
16. Container

Prob. 51-8. How many lead cells, each having an emf of 2 volts and an internal resistance of 0.005 ohm, will be needed to supply ten incandescent lamps in parallel, if each lamp requires 0.5 ampere at 32 volts?

Prob. 52-8. A battery of fifteen lead cells is to be charged in series from a 110-volt d-c line. Each cell has an emf of 2.02 volts and an internal resistance of 0.006 ohm. What size of resistor, measured in ohms, must be placed in series with the cell so that the charging current will not exceed 25 amperes?

Prob. 53-8. A set of eighty lead-acid storage cells, each having 2.02 volts as its internal emf and an internal resistance of 0.005 ohm, is to be charged in two parallel groupings having forty cells in series in each group. The normal charging current for each cell is 40 amperes. What must be the terminal voltage of the generator supplying the charging current?

Prob. 54-8. If all eighty of the cells of Prob. 53-8 are discharged in series at the 40-ampere rate, what is the terminal voltage of the battery?

Prob. 55-8. A set of ninety-six lead storage cells, each having 2 volts emf and 0.004 ohm of internal resistance, is to be charged from d-c line wires between which the potential is maintained constant at 110 volts. If each cell is to be charged at a 20-ampere rate, how should these cells be arranged in order to have the least power lost in a series resistance connected between the battery to be charged and the line wires?

Prob. 56-8. How could the cells of Prob. 55-8 be arranged in order to deliver 160 amperes to a load, without exceeding a current of 20 amperes in any one cell? What would the terminal voltage be for the battery thus arranged?

Prob. 57-8. Each cell of a 6-volt automotive battery has an emf of 2.1 volts and an internal resistance of 0.002 ohm. What is the open-circuit voltage of the battery?

Prob. 58-8. What is the terminal voltage of the battery of Prob. 57-8 if the starting motor requires a current of 200 amperes?

Prob. 59-8. If six cells having the characteristics indicated in Prob. 57-8 are connected to form a 12-volt automotive battery, what is its terminal voltage when it supplies a current of 300 amperes?

Prob. 60-8. In nighttime driving, the lights and ignition of an automobile may require a current of 14 amperes from the 12-volt battery. If the battery has the characteristics of the battery of Prob. 59-8, what is its terminal voltage when supplying this nighttime load?

17. Rating of Batteries. Whether a battery cell is of the primary or of the storage type, it is helpful to have a specification of the length of time the cell will be able to deliver a useful current. Since current is the rate in coulombs per second that charge is removed from the cell, charge could be expressed in terms of current multiplied by time, or amperes × seconds = coulombs. However, custom has dictated a larger unit of measure of battery charge, called the *ampere-hour.* One ampere-hour means the total charge that would be delivered by a current of 1 ampere acting throughout 1 hour, or 3600 coulombs. For dry cells and other small batteries for operating electronic equipment, the ampere-hour is too large; the smaller unit used to describe the charge capacity of such small batteries is the *milliampere-hour,* or the charge carried from the battery by 1 milliampere acting throughout 1 hour, i.e., 3.6 coulombs.

However, the chemical actions inside a battery do not permit a simple listing of the ampere-hour capacity of a battery. The internal resistance of a cell changes in several different ways as the cell is discharged. If a large current is taken from a small battery, the increase in internal resistance accompanying the chemical changes may be so large that the battery may become exhausted in a very short time. On the other hand, if the same battery is required to supply only a small current, the internal chemical changes may be utilized in such a way that the total charge (in ampere-hours) supplied by the battery is considerably larger than the capacity it had for the fast discharge rate.

These effects of internal changes on the amount of charge that can be drawn out of a battery are expressed in terms of a *time duration* stated with the ampere-hour specification. For example, a small cell for use in a satellite is rated at 20 milliampere-hours for a 10-hour discharge time. An automotive cell might be rated at 100 ampere-hours for a 20-hour discharge time. Each of these figures taken together specifies a "normal" discharge current for the cell.

Example 22. What are the normal discharge currents of the two cells listed above?

The normal discharge current is found by *dividing* the ampere-hour rating by the time duration specified. For the satellite cell, the normal discharge current is $\frac{20}{10} = 2$ *milliamperes*. For the automotive cell, the normal current is $\frac{100}{20} = 5$ *amperes*.

When the capacity of a battery is expressed in terms of its ampere-hour rating, the description primarily tells about the ability of a battery to sustain a reasonable load for a relatively long period of time. However, many tasks of space flight are similar to the cranking task of an automotive battery, i.e., an extremely large current must be supplied for a short time, usually a few seconds. Because the 10-hour and 20-hour durations are not useful for such specification, two other figures-of-merit are commonly stated for these short-time operations. A large current is chosen first, usually 300 amperes for automotive batteries. Then the cranking ability is specified (1) as the terminal voltage that occurs in a given number of seconds after this excessive discharge is started, and (2) the number of minutes that the battery can keep supplying this fixed current until its terminal voltage becomes uselessly low, perhaps only half of its nominal value.

Example 23. A 6-volt automotive battery with a 20-hour rating of 100 ampere-hours has a 5-second voltage of 4.4 volts for a current of 300 amperes, and at 300 amperes the terminal voltage becomes 1.0 volt per cell after 4.5 minutes. Explain these ratings.

If the battery were discharged at its 20-hour rating, the current would be 5 amperes, as in Example 22. The endpoint of this "normal" rating is taken as a terminal voltage of 5.25 volts, according to the battery manufacturers' standards. If the cell starts at 6.3 volts on open circuit (2.1 volts per cell) and is discharged at the normal current of 5 amperes, the terminal voltage (with 5 amperes still being supplied) will become 5.25 volts at the end of 20 hours, or 1.75 volts per cell.

However, if the fully charged battery again has an open-circuit voltage of 6.3 volts, the voltage drops sharply in 5 seconds of supplying 300 amperes, to become only 4.4 volts (measured while the current is still being supplied). If the current is still maintained at 300 amperes by continual reduction of the load resistance, the terminal voltage continues to drop. Although the battery could function successfully for 20 hours for the normal current drain of 5 amperes, its terminal voltage drops drastically to only 3.0 volts (or 1.0 volt per cell) in just 4.5 minutes for the heavy cranking drain.

In the sections concerning arrangements of cells into batteries, the emf of each cell and its internal resistance were considered constant. However, the internal chemical reactions that accompany the charging and discharging of lead-acid batteries do result in changes in *both* internal emf and internal resistance. Polarization by gases may bring about rather considerable changes of both values, and such changes may be temporary. After a waiting period, new values of emf and internal resistance may appear. Because of these variations, the recommendations of the manufacturer of the battery should be followed closely in operating a storage battery. Also the instructions concerning any battery-charging apparatus should similarly be considered carefully so that damage does not result either to the battery or to the charging equipment.

A meter of the type shown in Fig. 150 may be used to indicate the number of ampere-hours of charge that have been taken from a battery, particularly when the battery is to be operated through many cycles of charge and discharge. The ampere-hour meter functions exactly in the same way as does the watthour meter of Fig. 117, except that the meter totalizes amperes over time in hours, rather than watts over time in hours. To produce an operation of the disc in a mercury pool independent of voltage, permanent magnets are used in place of the potential coils of the watthour meter. Since permanent magnets act on a conducting disc in this form of ampere-hour meter to produce the turning effort, the same disc also develops the necessary braking drag for stable indications. Auxiliary circuitry associated with an ampere-

Fig. 150. Mercury-motor ampere-hour meter. *Sangamo Electric Company.*

hour meter can be arranged to start and stop a battery charger automatically when preset amounts of charge in ampere-hours have been taken out of or added to a battery.

Several types of storage battery are often shipped in what is called the *dry-charged* state. The battery is manufactured in such a way that the positive and negative plates are filled with the materials that would be found in a fully charged cell, but the cells are sealed without any electrolyte in them. When the battery is to be installed, premixed electrolyte is added to the cells. The battery will be able to supply considerable charge immediately, or it can be brought to its full ampere-hour capacity with just a short period of charging.

18. Care of Automotive Batteries.

(1) Appropriate settings for the generator regulator should be made to assure a reasonable state of charge in the battery at all times.

(2) Driving habits and the use of accessories are important factors in keeping a battery in a satisfactory state of charge.

(3) Low-water consumption by the battery is an indication of the proper setting for the generator regulator.

(4) The specific gravity of a cell is another indicator of its state of

charge: current practice chooses electrolyte with a specific gravity of 1.260 for full charge, thereby yielding a figure of 1.070 for complete discharge at the 20-hour rate.

(5) Unless electrolyte has been lost through spilling or leaking, it should not be necessary to add any acid to a battery during its lifetime.

(6) When separate charging of a battery becomes necessary, there are three methods to consider: *constant-current* charging, *constant-potential* charging, and *high-rate* charging.

(7) In charging, connect the positive terminal of the source of charging power to the positive terminal of the battery and set the current according to the scheme being used; the charger must furnish unidirectional current (d-c) to the battery.

(8) *Constant-current* charging is often performed at a rate of 1 ampere per positive plate in each cell: for a six-cell, 12-volt battery having seven positive plates in each cell, the charge rate would be 7 amperes.

(9) *Constant-potential* charging causes the charging current to taper off slowly as charge accumulates in the battery, providing a satisfactory charge if the initial current is not too large for the battery.

(10) *High-rate* charging is usually a form of constant-potential charging in which a large-capacity machine supplies 40 to 50 amperes as the initial current for a 12-volt battery.

(11) When a battery must remain unused for a long time, it should be checked periodically and brought up to full charge again whenever a drop in specific gravity to 1.220 is observed.

(12) Since the mixture of hydrogen and oxygen gases issuing from the top of battery cells on charge is highly explosive, do not bring a flame or sparks near the vent openings.

(13) Since the sulfuric acid in battery electrolyte will destroy clothing and human tissue, great care should be exercised to avoid any contact; if electrolyte is spattered into the eyes, wash it out immediately with plenty of clear cold water, and seek medical attention if this first-aid procedure does not relieve the discomfort.

19. Specialized Industrial Batteries. When operating in an automobile, the usual lead-acid battery is normally kept reasonably well charged through the action of the generator and the generator regulator, with the battery furnishing power just during brief starting periods and occasionally to supply lights and accessories when the gasoline engine is not operating. For industrial use, lead-acid batteries furnish power for materials-handling trucks and lifts as the only source of power. In such service, the battery must undergo a "deep" discharge during working hours; then charging in idle hours returns the battery to useful

8: Batteries

working condition again. Industrial batteries for supplying power for emergency operations in telephone and power plants may be operated either in the "floating" condition of the automotive battery, or in the "cycling" condition of the industrial-truck service. However, the considerations of size, weight, lifetime, and cost lead to industrial designs that differ from the automotive battery, for either kind of charging.

The active material pasted into the flat positive plates of automotive batteries would shed away after a few cycles of deep-discharge service. For *industrial-truck batteries,* a rugged positive plate has been evolved. The construction shown in Fig. 151 indicates how the active materials are enclosed in long tubes of woven glass fiber, supported on long grid

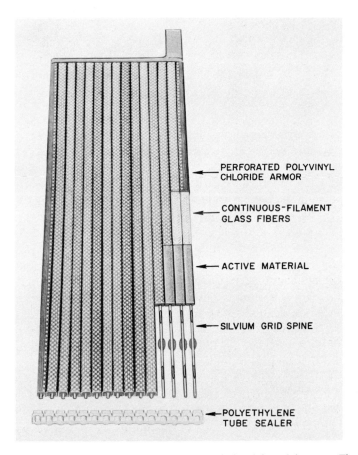

Fig. 151. Square-tube positive-plate construction for industrial-truck battery. *The Electric Storage Battery Company.*

168 Essentials of Electricity

Fig. 152. Fifteen-cell truck battery. *The Electric Storage Battery Company.*

forms of acid-resisting alloy. Into such a positive plate the electrolyte may penetrate quickly to sustain sudden heavy discharges, while the active material is kept in place on the electrode despite heating and the accompanying forces of expansion. Other refinements in construction make possible the battery shown in Fig. 152, each cell having a 6-hour service rating of 425 ampere-hours. With a standard height of $22\frac{5}{8}$ inches, this battery of fifteen cells is about 3 feet long and 16 inches wide.

For railway-lighting and other similar loads, there are occasions when the battery may be kept floating on a generator. At other times, the battery may have to carry the entire load for several hours. For such auxiliary service, a storage battery invented by Thomas Edison finds worthwhile application. The construction, shown in Fig. 153, is considerably different from that of a lead-acid battery. Nickel oxide and nickel flakes are packed together into nickeled-steel tubes on nickeled-steel grids to form positive plates. Iron oxide is packed into rectangular openings in the negative plates. When these electrodes are suitably charged in an electrolyte of a solution of potassium hydroxide in water, the nickel-alkaline cell produces an emf of about 1.3 volts, with a terminal voltage of about 1.2 volts in operation. A special advantage of the nickel-alkaline battery is that the specific gravity of the electrolyte does not change during charge and discharge. The container of welded nickeled steel keeps the materials of the cell well protected against any contamination, and within wide limits the cell can withstand abuse that would quickly end the life of other types of storage battery. A disadvantage of the steel container is that it must be insulated from the container of adjacent cells. Such insulation is usually provided by a mounting crate made of wood that keeps the individual cells of the battery separated. A steel cradle may be built around the insulated

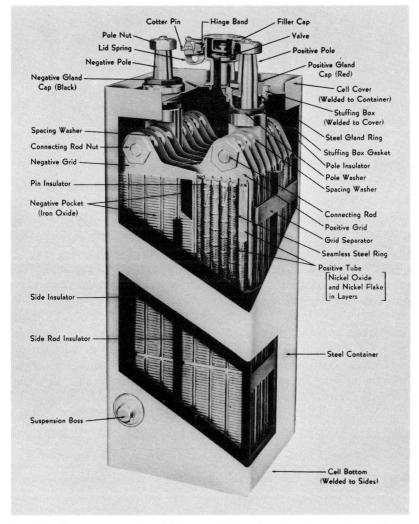

Fig. 153. Cutaway view of a nickel-iron-alkaline cell. *The Electric Storage Battery Company.*

170 Essentials of Electricity

Fig. 154. Subway-car battery for 32-volt emergency control service. *The Electric Storage Battery Company.*

support for safe lifting of the battery. Figure 154 shows a battery in place under the seat of a subway car, to provide auxiliary lighting when the normal power from the third rail is disconnected.

Many storage batteries cannot supply large power for periods of time longer than those normally encountered in starting an automotive engine. A modification of the nickel-alkaline cell known as a *nickel-cadmium cell* has the desirable properties of sustaining heavy discharges for a considerable length of time. In addition to adding cadmium to the iron of the negative electrode, the plates of the nickel-cadmium cell contain the active materials in very small pockets into which the electrolyte may penetrate readily to sustain the current of high-rate discharges. The battery of six cells shown in Fig. 155 can sustain a discharge of 300 amperes for 10 minutes with only approximately 20% drop in terminal voltage. Furthermore, the cells of Fig. 155 can supply a current of 1900 amperes for 30 seconds, after which the terminal voltage is only half of its original open-circuit value of 1.3 volts per cell. Such capacity finds use in the cranking of diesel engines, which normally require a longer cranking period than comparable gasoline engines, especially in railroad locomotives in which extremes of temperature may be encountered.

Sealed nickel-cadmium cells manufactured in wafer form offer many of the features of ruggedness and ease of charging of the larger types,

Fig. 155. One section of a diesel-cranking locomotive battery. *The Electric Storage Battery Company.*

with the complete portability of ordinary dry cells, yet rechargeable. Figure 156 shows a 6-volt battery assembled from nickel-cadmium cells, offering 225 milliampere-hours at a 10-hour discharge rate. Such batteries perform useful functions in operating equipment needed in space flight.

A *silver-zinc cell* also employing potassium hydroxide as its electrolyte has been developed for missile and space supplies. In a cold environment, the electrolyte may be kept in a special container outside the cell, to be inserted automatically at the appropriate moment of flight. Other automatic equipment heats the electrolyte in a chemically warmed radiator while the cell is being filled. This process permits the dry-charged silver-zinc primary cell to deliver its full

Fig. 156. Five sealed wafers make up this small rechargeable battery. *Burgess Battery Company.*

capacity at the moment required. Storage cells of the silver-zinc construction afford large ampere-hour capacity in small sizes for the short-duration rating needed.

Whereas the forms of battery previously discussed convert chemical energy into electric form, two other forms of electric cell are also available for specialized applications. The energy of sunlight is converted to d-c power in *solar cells,* and the energy of decomposing radioactive materials provides suitable voltage and current in *nuclear cells.*

SUMMARY

An ELECTRIC BATTERY transforms chemical energy into electric energy. It consists of two unlike conductors called positive and negative plates immersed in a fluid which attacks one of the plates chemically. The voltage set up by this chemical action is called the electromotive force, commonly written *emf.*

A CELL of a battery is characterized by its OPEN-CIRCUIT VOLTAGE (or emf) and its INTERNAL RESISTANCE.

When a cell supplies current to a load, the TERMINAL VOLTAGE of the cell drops below its emf because of the drop in the internal resistance.

Combinations of cells may be reduced to a single linear equivalent circuit for the battery, with the aid of the THEVENIN METHOD.

For a FIXED NUMBER OF CELLS, the largest current will be supplied to a given load when the cells are connected into a series-parallel arrangement having its Thevenin internal resistance as nearly as possible equal to the resistance of the load.

DRY CELLS are usually of the carbon-zinc type, with ammonium chloride as an electrolyte in a paste, the action of which sets up an emf of about 1.5 volts.

In ordinary dry cells, the ZINC IS CONSUMED AS A FUEL. Its high cost prevents extensive use of such batteries as a source of large quantities of electric power.

LOCAL ACTION occurs in a battery cell when impurities form a sort of localized small cell that discharges the main cell uselessly and wastefully.

POLARIZATION is the process of forming bubbles of hydrogen gas on the positive plate of a cell, altering the internal behavior of the cell so that little terminal voltage can be obtained unless the bubbles are removed.

If small dry cells are to be TESTED, terminal voltage should be measured when each cell is supplying a known safe current.

ELECTROLYSIS is the opposite of the battery effect. When a current is sent through a metal salt solution, it takes the metal out of solution and deposits it on the negative plate of the electrolytic cell. The weight of metal deposited is always the same per ampere-hour for the same solution. In a copper sulfate solution, copper may be deposited on the negative electrode for electroplating or electrotyping. Silver plating may be produced similarly in a solution of silver nitrate.

ELECTROLYTIC DAMAGE TO WATER AND GAS MAINS takes place wherever a current, already established in the pipe walls, leaves the pipe through moist earth to go to some other conductor.

STORAGE BATTERIES change their electrode structure in a useful form of electrolysis during charge. Electric energy is stored in chemical form, to be released upon discharge.

AUTOMOTIVE BATTERIES have lead and lead peroxide as the active materials for their electrodes, in a solution of sulfuric acid, yielding an emf of 2.10 volts per cell.

RATING of the charge-handling capacity of a battery is generally specified in ampere-hours for a specified duration of time of discharge.

Some rules for the CARE of automotive and other lead-acid batteries are given in Section 8-18.

RUGGEDIZED forms of lead-acid batteries are needed for industrial-truck service and other deep-discharge loads.

NICKEL-ALKALINE CELLS of various forms offer advantages of easy maintenance and sustained power for cranking diesel engines or providing railway lighting.

FOR SPACE FLIGHT, specialized cells are designed to fit the requirements of the instruments to be supplied.

PROBLEMS

Prob. 61-8. Five cells are available, each having an emf of 1.48 volts and an internal resistance of 0.4 ohm. What current will the series connection of these cells establish in a circuit having a resistance of 12 ohms?

Prob. 62-8. Connected in series are 56 storage cells, each having an emf of 2.1 volts and internal resistance of 0.015 ohm, to supply a paralleled bank of fifteen incandescent lamps, each of which has a resistance of 200 ohms. The resistance of the connecting wires may be neglected.
(a) What is the current in the line wires from the battery?
(b) What is the current in each lamp?

Prob. 63-8. If the line wires of Prob. 62-8 have a total resistance of 0.6 ohm, what will be the current from the battery to the lamps?

Prob. 64-8. What is the terminal voltage of each cell (a) in Prob. 62-8, (b) in Prob. 63-8?

Prob. 65-8. What is the voltage across each lamp in Prob. 63-8?

Prob. 66-8. What power is lost in the line in Prob. 63-8?

Prob. 67-8. What is the total power lost in the cells in Prob. 63-8?

Prob. 68-8. The terminal potential of a certain cell is 1.2 volts when delivering 2.2 amperes. When the cell is delivering 3 amperes, the terminal voltage is 0.9 volt. Draw the graph of terminal voltage versus current delivered for this cell.

Prob. 69-8. What is the internal resistance of the cell of Prob. 68-8?

Prob. 70-8. What is the internal emf of the cell of Prob. 68-8?

Prob. 71-8. What is the short-circuit current of the cell of Prob. 68-8?

Prob. 72-8. A telegraph line is to have a current of 0.25 ampere through its total resistance of 100 ohms. How many cells, each having an emf of 1.08 volts and an internal resistance of 2 ohms would be required to operate the line under these conditions?

Prob. 73-8. A battery of six dry cells in series is connected to a load having a resistance of 0.08 ohm. Each cell has an emf of 1.5 volts and an internal resistance of 0.05 ohm. What is the load current?

Prob. 74-8. If the cells in Prob. 73-8 were arranged in parallel, what would the load current be?

Prob. 75-8. If the cells in Prob. 73-8 were arranged in two paralleled sets of three cells in series, what would the load current be?

Prob. 76-8. Find the Thevenin equivalent circuit for the battery: (*a*) in Prob. 73-8, (*b*) in Prob. 74-8, and (*c*) in Prob. 75-8.

Prob. 77-8. Determine the current in each cell: (*a*) in Prob. 73-8, (*b*) in Prob. 74-8, and (*c*) in Prob. 75-8.

Prob. 78-8. What current would be created in the load resistance if the cells of Prob. 73-8 were arranged in three paralleled sets of two cells in series?

Prob. 79-8. Draw the graph of terminal voltage versus current for the battery of cells arranged as in Prob. 78-8.

Prob. 80-8. The normal current of each storage cell in a large battery is 5 amperes, with an internal resistance of 0.002 ohm and an emf of 2.05 volts. The battery is arranged to light 40 paralleled lamps, each of which requires 0.5 ampere at 112 volts. The connecting wires have negligible resistance. How many cells are needed in this large battery, and how are they arranged?

Prob. 81-8. If each of the two line wires for Prob. 80-8 has a resistance of 0.3 ohm, how many cells must the battery have and what must be their arrangement to satisfy the same conditions?

Prob. 82-8. What is the terminal voltage of the battery in Prob. 81-8?

Prob. 83-8. A generator delivers 120 amperes at 115 volts. In case of accident to the generator, a battery of storage cells is kept as a reserve supply. Each cell has an emf of 2.10 volts and an internal resistance of 0.0015 ohm, and an 8-hour rating of 320 ampere-hours. If the battery is large enough to take the place of the generator for 1 hour (assume that the 1-hour capacity is only 160 ampere-hours), how many cells should be used and how should they be arranged?

Prob. 84-8. If the battery of cells arranged as in Prob. 83-8 are charged with the current indicated by their 8-hour rating, what charging voltage is needed?

Prob. 85-8. For the battery of cells connected as in Prob. 80-8, what charging voltage is needed when the charging current is 5 amperes?

Prob. 86-8. An inexpensive 5-volt voltmeter with a sensitivity of only 10 ohms per volt is attached to a 4.5-volt battery of dry cells having very little capacity. Although the emf of the battery is 4.5 volts, the voltmeter indicates only 4.2 volts.

(*a*) What current does the voltmeter take from the battery?

(*b*) What is the internal resistance of the battery?

Prob. 87-8. If a 5-volt voltmeter with a sensitivity of 20 ohms per volt were used for the measurement in Prob. 86-8, what voltage would it indicate?

Prob. 88-8. If a 6-volt voltmeter having 100 ohms per volt as its figure-of-merit is used for the measurement in Prob. 86-8, what will it read? How much current will this voltmeter require from the battery?

Prob. 89-8. A certain dry cell has an emf of 1.5 volts and an internal resistance of 0.3 ohm.

(*a*) If a suitably large ammeter with an internal resistance of 0.1 ohm is momentarily connected directly across the terminals of the cell, what current will it indicate?

(*b*) What current will this cell establish if a 0.3-ohm resistor is placed directly across its terminals?

(*c*) What is the terminal voltage of the cell, in part (*a*), and in part (*b*)?

Prob. 90-8. When operated from a storage battery with a terminal potential of 6 volts, the headlights of a certain automobile require 10 amperes. In an

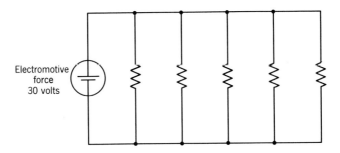

Fig. 157. The resistance of the battery is important.

emergency a man attempted to supply these lamps from four of the cells of Prob. 89-8 connected in series, after he had disconnected the run-down storage battery. What current was established in the headlights?

Prob. 91-8. What is the terminal voltage of the four-cell battery of Prob. 90-8?

Prob. 92-8. How many of the cells of Prob. 89-8 must be connected in series to light the lamps of Prob. 90-8 at normal brightness?

Prob. 93-8. Repeat Prob. 90-8 with cells having an emf of 1.5 volts and a low internal resistance of 0.05 ohm.

Prob. 94-8. Repeat Prob. 91-8 with the cells of Prob. 93-8.

Prob. 95-8. Repeat Prob. 92-8 with the cells of Prob. 93-8.

Prob. 96-8. How many hours would a 5-pound plate of silver last in a plating operation in which the average current is 10 amperes?

Prob. 97-8. What would be the length of time in Prob. 96-8 if the plate were copper instead of silver?

Prob. 98-8. In Fig. 157, each of the load resistors has a resistance of 5 ohms. The battery has an emf of 30 volts and an internal resistance of 1.0 ohm. What is the current supplied to each resistor?

Prob. 99-8. If one of the resistors is disconnected from the line wires in Prob. 98-8, what current is supplied to each of the remaining resistors?

Prob. 100-8. If three of the load resistors of Prob. 98-8 are disconnected from the line wires, what current is supplied to each of the remaining pair?

Prob. 101-8. Repeat Prob. 98-8 if the battery has an emf of 30 volts with a lower internal resistance of 0.03 ohm.

Prob. 102-8. Repeat Prob. 99-8 for the battery of Prob. 101-8.

Prob. 103-8. Repeat Prob. 100-8 for the battery of Prob. 101-8.

Prob. 104-8. Review the results of Probs. 96-8 through 103-8. Explain why a source of low internal resistance is needed if several paralleled loads are to be turned on and off independently of each other.

Prob. 105-8. Use the Thevenin method to determine the voltage across the 24-ohm load in Fig. 99 of Chapter 6.

Prob. 106-8. Find the current in the 40-ohm shunt arm of the circuit of Fig. 99 in Chapter 6 through application of the Thevenin method.

Prob. 107-8. By means of the Thevenin method, find the current in the 16-ohm series arm of the ladder circuit of Fig. 99 in Chapter 6.

9

Wire and Wiring Systems

1. Insulators and Conductors. Some substances conduct electricity more readily than do others. Therefore, it is customary to consider one material as an insulator if a current can only be established in it with a very high voltage, while another material may be considered a conductor if a small voltage can set up a relatively large current in it.

The more common insulators are porcelain, glass, mica, cellulose acetate (under various trade names), polystyrene, polyethylene, synthetic rubber, jute, and paper. The last five named insulators are employed extensively in the construction of electric power and communication cables.

The best conductors in their order of decreasing resistance are four metals: silver, copper, gold, and aluminum. Of these, the cheaper price for copper and aluminum commends them to general use as electrical conductors. These two are in sharp competition in the field of power transmission lines, but copper is used almost exclusively for interior wiring due to its greater toughness and smaller size for wires of a given current-carrying capacity. Aluminum may create a somewhat greater fire hazard because of its lower melting temperature.

In Chapters 6 and 7 the effects of resistance in line wires on the behavior of circuits was studied. Since the choice of a given type of wire thereby selects the resistance introduced into the circuit by this wire, it is important to know how resistance changes with the dimensions of wire. In order to find out how to compute the resistance of a piece of wire, it will be interesting to note separately the effects of length and diameter.

2. Resistance Increases With Length. A round copper wire 0.1 inch in diameter has a resistance of about 1 ohm for a 1000-foot length. The resistance of 2000 feet of this wire would have twice as much resistance as 1000 feet, or 2 ohms, since the two 1-ohm resistances of each 1000-foot length are in series. Similarly, the resistance of 10,000 feet of this wire would be ten times the resistance of 1000 feet. The resistance of a given size of wire is thus seen to be the resistance of 1000 feet multiplied by the number of thousand-foot lengths, or it is the resistance of 1 foot multiplied by the number of feet.

Example 1. Copper wire of a certain size has a resistance of 2.57 ohms per 1000 feet. What is the resistance of 400 feet of this wire?

resistance of 1000 feet = 2.57 ohms
resistance of 400 feet = 0.4 × 2.57
= 1.03 ohms, rounded off

Prob. 1-9. Wire of the size stated in Example 1 is available in long lengths. (a) What is the resistance of a mile of this wire? (b) What would be the resistance of a 2-mile two-wire line of this wire?

Prob. 2-9. Wire sometimes used in exterior wiring has a resistance of 1.586 ohms per 1000 feet. What is the resistance of a two-wire 250-foot lead-in from the street to a well setback house if this wire is chosen?

Prob. 3-9. The resistance of 1 foot of a certain size wire is 0.083 ohm. How many feet of this wire will it take to make a resistance of 30 ohms?

Prob. 4-9. If the resistance of 900 feet of a certain size wire is 0.223 ohm, what is the resistance of 1500 feet?

3. Resistance Decreases as Area Increases. The effect of the cross-sectional area of a wire on resistance is more complicated than the effect of length. Let us assume for a moment that copper wires are drawn with a square cross section, instead of being round. Assume a certain square wire to measure 1 inch on each side. The cross-sectional area of the wire is then a square of 1 inch by 1 inch, i.e., it is an area of 1 square inch.

If we take a square wire 2 inches on each side, its cross-sectional area would be 2 × 2 = 4 square inches. In Fig. 158 (a), an end view of a square wire 1 inch on a side is shown. Figure 158 (b) shows the end view of a square wire measuring 2 inches on a side. Note that although the length of one side of the wire shown in Fig. 158 (b) is only *twice* as great as the length of one side of the wire shown in Fig. 158 (a), the area of the wire in (b) is *four times* as great as the area of the wire in (a). In other words, four 1-inch wires could be made out of a single

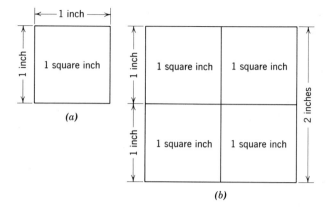

Fig. 158. Although the side of the large square of (b) is twice the side of (a), the area of (b) is four times the area of (a).

2-inch wire of square cross-section. Figure 158 shows clearly how the material in the 2-inch wire could be divided into four 1-inch wires.

In the same way it will be seen that if we make a square wire 3 inches on a side, the area of the cross section will be $3 \times 3 = 9$ square inches. A square wire 3 inches on a side could thus be divided into nine square wires 1 inch on a side. Similarly, a 5-inch wire could be divided into 25 wires 1 inch on a side, etc. For each of the wires listed above, note that the cross-sectional area of the end increases as the *square of the number of times* the length of the side of the wire is increased, and that the number of small wires that are contained in the larger one (or equivalent to the larger one) is also equal to the *square* of the number of times the length of the side is increased. Thus:

(a) A 2-inch wire has a 4 square-inch area and is equal to four 1-inch wires.

(b) A 3-inch wire has a 9 square-inch area and is equal to nine 1-inch wires.

(c) A 4-inch wire has a 16 square-inch area and is equal to 16 1-inch wires.

Measurement on a square copper wire 1 foot long and $\frac{1}{100}$ inch on a side shows that its resistance is 0.081 ohm. If a second square copper wire 1 foot long was $\frac{2}{100}$ inch on a side, as shown in Fig. 159, this $\frac{2}{100}$-inch wire is equivalent to four $\frac{1}{100}$-inch wires 1 foot long laid side by side. If connections were made to the ends of the four wires, it is clear that these four wires are essentially in parallel. Thus the resistance of 1 foot of $\frac{2}{100}$-inch wire is equal to the resistance of four $\frac{1}{100}$-inch wires each 1 foot long and connected in parallel.

For a given voltage, the current set up in the larger wire would then be *four times* as large as the current in the smaller wire. In other words, the *resistance* of the larger wire is only *one-fourth* the resistance of the smaller wire. Recall from Chapter 3 that the equivalent resistance of four paralleled equal resistances is one-fourth of the resistance of one of them. Therefore, if the resistance of 1 foot of $\frac{1}{100}$-inch square copper wire is 0.081 ohm as noted above, the resistance of 1 foot of $\frac{2}{100}$-inch square copper wire would be $(0.081/4) = 0.0203$ ohm. Similarly, the resistance of 1 foot of $\frac{3}{100}$-inch square copper wire would be $(0.081/9) = 0.009$ ohm.

Note that if we know the resistance of a foot of any square wire of unit length of side, we can find the resistance of a foot of square wire having any length of its side by *dividing* the resistance of the unit wire by the *square* of the number of unit lengths contained in the side of the larger wire. This process for finding the resistance of a larger wire is the same as *dividing* the resistance of the unit wire by the *number* of unit wires contained in the larger wire.

Example 2. Assume that a certain square wire, $\frac{1}{1000}$ inch on a side and 1 foot long, has a resistance of 8.16 ohms. What would be the resistance of a square wire 1 foot long and $\frac{8}{1000}$ inch on a side?

Since each wire is just 1 foot long, their resistances differ because of their different cross sections:

$$\text{resistance of wire } \tfrac{1}{1000} \text{ inch on a side} = 8.16 \text{ ohms}$$

$$\text{resistance of wire } \tfrac{8}{1000} \text{ inch on a side} = \frac{8.16}{(8)^2} = \frac{8.16}{64}$$

$$= 0.127 \text{ ohm}$$

A wire that is $\frac{8}{1000}$ inch on a side is equivalent to $8 \times 8 = 64$ wires $\frac{1}{1000}$ inch on a side, all laid side by side, i.e., in parallel. The resistance of an $\frac{8}{1000}$-inch wire therefore equals $\frac{1}{64}$ of 8.16 ohms, or 0.127 ohm.

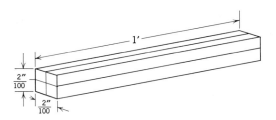

Fig. 159. One wire 2 inches square is equivalent to four wires 1 inch square.

Example 3. Given the 8.16-ohm resistance of the 1-foot length of square wire $\frac{1}{1000}$ inch on a side, as in Example 2, what would be the resistance of a square wire 0.415 inch on a side?

$$0.415 = \frac{415}{1000}$$

Thus, this large wire would be equal to $415 \times 415 = 172{,}000$ of the $\frac{1}{1000}$-inch wires laid side by side, or in parallel.

Resistance of 0.415-inch wire then equals

$$\frac{8.16}{172{,}000} = 0.0000474 \text{ ohm}$$

With the numbers arranged as in Example 2:

resistance of wire $\frac{1}{1000}$ inch on a side = 8.16 ohms

resistance of wire $\frac{415}{1000}$ inch on a side = $\dfrac{8.16}{415 \times 415}$

= 0.0000474 ohm

Prob. 5-9. Given the resistance of a unit square wire as in Example 2, what is the resistance of 1 foot of square wire 0.045 inch on a side?

Prob. 6-9. Assuming that the wire of Example 2 is copper, what is the resistance of 1 foot of square copper wire 0.5 inch on a side?

Prob. 7-9. Given the unit copper wire of Example 2, what must be the length of side of 1 foot of square copper wire if its resistance is to be 0.0835 ohm?

4. Circular Wire. Because most copper wire is drawn with a cross section that is round instead of square, we must also be able to compute the resistance of round wires. It may be asserted that the statements concerning square wire apply equally to round wire. In particular:

(1) The resistance of any length of round wires is equal to the resistance of 1 foot of that wire multiplied by the length in feet.

(2) If the resistance of 1 foot of round copper wire $\frac{1}{1000}$ inch in diameter is known, the resistance of 1 foot of any size of round wire is equal to the known resistance *divided* by the *square* of the number of thousandths in the diameter.

The first statement follows from the idea of a series connection. The second statement should be examined further for corroboration.

We have seen how a square wire 3 inches on a side is equal to $3 \times 3 = 9$ square wires 1 inch on a side, laid in parallel. Similarly, a round wire 3 inches in diameter is equal to $3 \times 3 = 9$ round wires 1 inch in diameter, laid side by side in parallel. As indicated in Fig. 160, it can be shown that the 3-inch circle has exactly the same area as the *nine* 1-inch circles combined, and will contain *nine* 1-inch circles. In Fig. 160(*b*) the parts projecting beyond the large circle are to be used up in filling in the chinks left inside the large circle. Therefore, the large circle can be thought of as being composed of small circles of unit

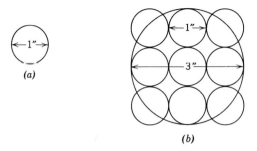

Fig. 160. Although the diameter of the large circle of (*b*) is three times the diameter of (*a*), the area of (*b*) is nine times the area of (*a*).

diameter. There are as many small circles in the large one as the square of the diameter of the large one.

Accordingly, a round wire $\frac{5}{1000}$ inch in diameter is equal to $5 \times 5 = 25$ round wires $\frac{1}{1000}$ inch in diameter laid side by side, or in parallel. If we knew the resistance of the $\frac{1}{1000}$-inch wire, we would know that the resistance of the $\frac{1}{1000}$-inch wire was $\frac{1}{25}$ of the resistance of the $\frac{1}{1000}$-inch wire.

The mathematical relations of the diameter of a circle to its area will be stated briefly here. The ratio of the *circumference* of a circle to its diameter is a constant number, designated by the Greek letter π, and equal to 3.1416 when rounded off. Through mathematical manipulation, the *area* of a circle can be shown to be $(\pi/4)$ times the *square* of the diameter of the circle. Here is another example of a proportion: The *area* of the circle is proportional to the *square* of its diameter. The important constant of proportionality in this ratio is $(\pi/4) = 0.7854$, also rounded off. Applying this ratio to the circles of Fig. 160 yields:

area of circle 1 inch in diam. = $0.7854 \times 1 \times 1 = 0.7854$ square inch
area of circle 3 inches in diam. = $0.7854 \times 3 \times 3 = 7.069$ square inches

There are then as many circles 1 inch in diameter in a circle 3 inches in diameter as:

0.7854 square inch is contained in 7.069 square inches, or 9

Therefore a circle 3 inches in diameter contains:

$3 \times 3 = 9$ circles 1 inch in diameter

In the same way it can be proved that a 5-inch circle contains $5 \times 5 = 25$ 1-inch circles, etc.

A round wire 1 foot long and $\frac{1}{1000}$ inch in diameter has been adopted as the unit round wire. The diameter of $\frac{1}{1000}$ inch is known as 1 *mil*. In the metric system, the prefix "milli-" means $\frac{1}{1000}$ of whatever the unit is that follows. In the coinage of the United States, we sometimes use the word "mill," meaning $\frac{1}{1000}$ of a dollar. When the spelling is *mil*, the meaning is always $\frac{1}{1000}$ *of an inch*. A wire of 1-mil in diameter is a wire $\frac{1}{1000}$ inch in diameter. A wire 25 mils in diameter is $\frac{25}{1000}$ inch in diameter, etc. Instead of saying "thousandths of an inch," we say "mils" in speaking of the diameter of wire.

Prob. 8-9. How many mils are there in $\frac{3}{4}$ of an inch?
Prob. 9-9. A wire has a diameter of 0.204 inch. What is its diameter in mils?
Prob. 10-9. What is the diameter in mils of a wire 0.460 inch in diameter?
Prob. 11-9. A large conductor for a power circuit is wrapped with paper until the thickness of the layer of paper is $\frac{1}{16}$ of an inch. What is the thickness of the insulation in mils?
Prob. 12-9. A wire has a diameter of 80.8 mils. What is its diameter in inches?

5. Circular Mil. Circular Mil-Foot. It has been shown in the preceding section that a wire of more than $\frac{1}{1000}$ inch or 1 mil in diameter will contain as many unit wires as the square of the number of mils or thousandths in its diameter. Thus, a wire 8 mils in diameter is equivalent to 64 wires 1 mil in diameter. Another way of stating this relation is to give a special name to the cross-sectional area of a wire having a 1-mil diameter. A round wire with a diameter of $\frac{1}{1000}$ inch or 1 mil is said to have a cross-sectional area of 1 *circular mil*, a usage intended to distinguish between square and circular cross sections. Expressed in this way, the end area of a wire $\frac{8}{1000}$ inch in diameter would be $8 \times 8 = 64$ circular mils.

Also noted in the preceding section was the fact that the unit wire for round wire is 1 foot long and 1 mil in diameter. Because the area of a circle 1 mil in diameter is called a circular mil, the unit round wire is called a *circular mil-foot*.

The cross-sectional area of a round wire expressed in circular mils is the number of unit circles which the area will contain, a unit circle being a circle 1 mil in diameter. Since any circle will contain as many unit circles as the square of the diameter in mils, we say that the area of a circle in circular mils equals the square of the diameter in mils. Thus:

the area of a circle 4 mils in diameter is $4 \times 4 = 16$ circular mils
area of a circle 60 mils in diameter is $60 \times 60 = 3600$ circular mils
area of a circle 250 mils in diameter is $250 \times 250 = 62,500$ circular mils

For calculations on round wires, this method of stating the area of a circle is much easier and more sensible than finding the area in square inches, a calculation that involves the value of π, or 3.1416, each time. This shorter method is merely finding the number of unit circles contained in a circle, instead of finding the number of unit squares. The number of unit circles (circular mils) is always the square of the diameter in mils.

Example 4. What is the area in circular mils of a circle $\frac{1}{2}$ inch in diameter?

$\frac{1}{2}$ inch = $\frac{500}{1000}$ inch = 500 mils
area = 500 × 500 = 250,000 circular mils

Example 5. How many unit wires would 1 foot of wire $\frac{1}{4}$ inch in diameter contain?

$\frac{1}{4}$ inch = $\frac{250}{1000}$ inch = 250 mils
area = 250 × 250 = 62,500 circular mils
area of unit wire = 1 circular mil

Wire of 62,500 circular mils then contains 62,500/1 = 62,500 unit wires.

Prob. 13-9. What is the circular-mil area of a wire 0.325 inch in diameter?
Prob. 14-9. What is the circular-mil area of wire 0.102 inch in diameter?
Prob. 15-9. Find the circular-mil area of a circle 0.003145 inch in diameter.
Prob. 16-9. How many unit wires will a wire 0.0285 inch in diameter make?
Prob. 17-9. To how many unit wires is a wire 1 inch in radius equivalent?
Prob. 18-9. What is the diameter of a wire containing 2500 circular mils?
Prob. 19-9. What is the diameter of a wire containing 16,900 circular mils?
Prob. 20-9. What area in circular mils will a circle 1.04 inches in diameter have?

6. Copper Wire. We have defined a unit wire as a round wire 1 foot long and 1 mil in diameter (i.e., with a cross-sectional area of 1 circular mil). This unit wire, called a circular mil-foot, when made of copper has a resistance of 10.4 ohms. Another way of stating this information is to say that copper has a *resistivity* of 10.4 ohms for a circular mil-foot, meaning that the resistance of a 1-foot length of wire having an area of 1 circular mil is equal to 10.4 ohms. Since the resistance of copper does change slightly with temperature, it is important to note that this resistance of 10.4 ohms for a circular mil-foot of copper wire is measured at 68° Fahrenheit, which is the same as 20° Centigrade.

This value of 10.4 ohms for a circular mil-foot of copper wire is important enough to be memorized. For if we know the resistance of 1 foot of round wire 1 mil in diameter, we can compute the resistance of 1 foot of any size round wire as follows:

To find the resistance of 1 foot of copper wire $\frac{6}{1000}$ inch in diameter, we first say that the resistance of 1 foot of copper wire having 1 mil diameter is 10.4 ohms. A wire having $\frac{6}{1000}$ inch, or 6 mils, diameter is equivalent to 6 × 6 = 36 wires 1 mil in diameter, laid side by side in parallel, since this larger wire contains $6 \times 6 = 36$ circular mils of area in its cross section. The resistance of 1 foot of 6-mil wire would be $10.4/36 = 0.289$ ohm, or:

resistance of 1 foot of wire 1 mil in diameter = 10.4 ohms

$$\text{resistance of 1 foot of wire 6 mils in diameter} = \frac{10.4}{6 \times 6}$$
$$= 0.289 \text{ ohm}$$

Furthermore, we have seen that if we know the resistance of 1 foot of any wire, we can find the resistance of any number of feet by simply multiplying by the length. Thus, if the resistance of 1 foot of copper wire 6 mils in diameter is 0.289 ohm, as found above, the resistance of 1000 feet of this wire would be $0.289 \times 1000 = 289$ ohms.

Therefore, to find the resistance of any length of any size wire:

Multiply the resistance of unit wire (1 circular mil-foot) by the length in feet and divide by the square of the mil diameter (circular-mil area).

Example 6. What is the resistance of 1 mile of copper wire 0.125 inch in diameter?

resistance of 1 foot of wire 1 mil in diameter = 10.4 ohms

$$\text{resistance of 1 foot of wire 125 mils in diameter} = \frac{10.4}{125 \times 125}$$
$$= 0.000667 \text{ ohm}$$

resistance of 5280 feet of wire 125 mils in diameter:
$$= 0.000667 \times 5280$$
$$= 3.52 \text{ ohms}$$

As a formula:

$$\text{resistance of wire} = \frac{\text{resistance per circular mil-foot} \times \text{length in feet}}{\text{circular-mil area}}$$
$$= \frac{10.4 \times 5280}{125 \times 125} = 3.52 \text{ ohms}$$

Prob. 21-9. How many feet of copper wire 0.04 inch in diameter will it take to make a resistance of 4 ohms?

Prob. 22-9. What will be the resistance of 900 feet of copper wire 0.25 inch in diameter?

Prob. 23-9. What diameter must copper wire have if 1 mile has a resistance of 0.196 ohm?

Prob. 24-9. What is the resistance of 2 miles of 0.32-mil copper wire?

Prob. 25-9. What is the circular-mil area of a wire $\frac{5}{16}$ inch in diameter?

Prob. 26-9. For copper wire of the size in Prob. 25-9, what will be the resistance of a 2000-foot length?

Prob. 27-9. What resistance has 525 feet of a copper wire of 9876 circular-mil area?

Prob. 28-9. What is the radius of the wire in Prob. 27-9?
Prob. 29-9. What is the resistance of 6 miles of copper wire having a diameter of $\frac{3}{8}$ inch?
Prob. 30-9. How many miles of copper wire $\frac{1}{2}$ inch in diameter will be required to make up a resistance of 3 ohms?
Prob. 31-9. The distance between a motor and a generator is 800 feet. Each of the two copper line wires has a diameter of 0.130 inch. What is the total resistance of the line?
Prob. 32-9. Ordinary lamp cord usually has a diameter of 0.040 inch. What is the resistance per 1000 feet?
Prob. 33-9. Doorbell wire often has a diameter of 0.040 inch. How many feet of such wire make up a resistance of 2 ohms?
Prob. 34-9. How many feet of the lamp cord in Prob. 32-9 would be needed to produce a resistance of 1 ohm?
Prob. 35-9. The line wires in Fig. 18 are each 400 feet long. Of what diameter must they be?
Prob. 36-9. How long is each line wire in Fig. 13 if each has a diameter of 0.064 inch?
Prob. 37-9. In Fig. 79 the distance from the generator to the lamps is 500 feet. What size wire is used?
Prob. 38-9. What is the resistance of 6 miles of copper wire $\frac{7}{16}$ inch in diameter?

7. Line Drop and Dimensions. If we know the length and size of a line wire and the current it is to carry, we can compute the voltage drop in that wire. If the return wire is of the same size and carries the same current, the line drop is just twice the voltage drop in one of the wires.

Example 7. A 2000-foot copper line wire is 0.204 inch in diameter. What is the voltage drop along the wire when the current through it is 40 amperes?

$$\text{resistance} = \frac{\text{resistance of unit wire} \times \text{length}}{\text{circular mils}}$$

$$= \frac{10.4 \times 2000}{204 \times 204} = 0.5 \text{ ohm}$$

$$\text{volts} = \text{ohms} \times \text{amperes}$$
$$= 0.5 \times 40 = 20 \text{ volts}$$

Prob. 39-9. What is the voltage drop along 400 feet of copper wire 0.064 inch in diameter when the current in the wire is 3 amperes?
Prob. 40-9. What voltage is required to establish a current of 20 amperes in a 1000-foot length of copper wire 0.162 inch in diameter?
Prob. 41-9. A $\frac{1}{4}$-inch copper wire carries 50 amperes. What is the voltage drop per mile?
Prob. 42-9. A 2500-foot length of copper wire carries 120 amperes. If its diameter is 0.364 inch, what is the voltage drop?
Prob. 43-9. What current in amperes can be established through 1500 feet of copper wire $\frac{1}{8}$ inch in diameter, if the voltage drop is to be 5 volts?
Prob. 44-9. What diameter in inches must a 1-mile length of copper wire have if a current of 25 amperes is to cause a drop of 6 volts in the wire?

Fig. 161. There is a drop in voltage along the line wires between generator G and motor M.

Prob. 45-9. If a potential of 12 volts is required to establish a current of 30 amperes in a line wire $\frac{3}{16}$ inch in diameter, what is the length of the line wire?

Prob. 46-9. The motor of Fig. 161 is 250 feet from the generator, and its current is 18 amperes. What size line wire must be used?

Prob. 47-9. If each of two copper line wires has a diameter of 0.262 inch, how far may a current of 25 amperes be transmitted with a line drop of only 8 volts?

Prob. 48-9. Each lamp in Fig. 162 requires 1 ampere at 115 volts. The lamps, 500 feet from the generator, are supplied over line wires each of which has a diameter of $\frac{1}{8}$ inch. What is the voltage of the generator?

Prob. 49-9. A copper wire is 500 feet long and 0.229 inch in diameter. What potential in volts will be required to set up a current of 15 amperes through the wire?

Prob. 50-9. What will be the drop per mile in a line consisting of copper wire $\frac{1}{10}$ inch in diameter carrying 20 amperes?

Prob. 51-9. A load is to receive 12 kilowatts at 550 volts, supplied over two copper wires each having a diameter of 0.162 inch. What is the line drop in volts per mile of line? What is the line loss in watts per mile?

Prob. 52-9. A group of incandescent lamps located 2500 feet from a generator requires 12 amperes. What size must each copper line wire be if the line drop is not to exceed 3.3 volts? State your answer as area in circular mils.

Prob. 53-9. A 220-volt 25-horsepower motor of 88% efficiency is located 500 feet from the generator, which supplies the motor over a copper-wire line, each conductor of which has a diameter of 0.460 inch. What is the voltage of the generator?

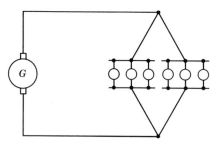

Fig. 162. The voltage at the generator must be higher than the voltage at the lamps.

Prob. 54-9. What size wire might have been used in Prob. 53-9 if a line drop of 3% of the voltage of the generator had been desired?

Prob. 55-9. A 115-volt generator and a 110-volt 10-horsepower motor of 85% efficiency are located 1800 feet apart. What size of copper wire is required between the two machines?

Prob. 56-9. What power is lost in a mile of line if the wire is 0.204 inch in diameter (for each conductor) and if the line is carrying 24 amperes?

Prob. 57-9. A building 2200 feet from a 115-volt generator is to be supplied with sufficient current from the generator to light 400 lamps in parallel, each lamp requiring 0.25 ampere. From the power generated, 4% is lost in the line wires. What circular-mil area is needed for each of two copper line wires for this installation?

8. Wire Table for Copper. In commercial practice, manufacturers make up only certain sizes of copper wire. These sizes are arranged according to a scale called a wire gage. In the United States, the scale that is standard is called the American Wire Gage (AWG), although sometimes it is designated by its earlier name of Brown and Sharpe Gage (B. & S.). Wire with diameters other than those of the AWG listing can be obtained from specialty manufacturers, usually at an increased cost.

The choice of diameters for the AWG numbers is essentially arbitrary, although the scheme roughly follows the number of passes needed to produce the given wire size. In particular, the sizes marked 0000 (sometimes called 4/0) and 30 were chosen to have diameters of 0.460 inch and 0.010 inch, respectively. Subdivision of the interval between these numbers was made by the mathematical relation known as geometrical progression, yielding a number by which the diameter of one size must be multiplied to give the next size larger. This multiple is 1.1229322, and from it the diameters that correspond to any of the gage numbers may be calculated. However, when the diameters are squared in the process of obtaining the circular-mil area of each wire size, different values can be obtained for the areas, depending on how the rounding off of the calculations is carried out.

In order to establish one set of numbers upon which American manufacturers could agree, the National Bureau of Standards carried out such calculations and arranged certain rules for rounding off numbers so that only one set of answers would be obtained. The resulting array of answers is known as a *wire table,* a device that greatly simplifies the computations needed for problems involving wires. Other values listed in wire tables are the resistance per 1000 feet of the given size of wire, and measures that relate weight to distance and resistance to weight. In the tables calculated by the National Bureau of Standards, the resistances are computed from a value of 10.371 ohms per circular mil-foot

188   Essentials of Electricity

and are rounded off according to the same set of rules as those used for dimensions.

An abridgement of the complete standard wire tables appears as Table A in the Appendix. The first column lists the AWG numbers, of which only the even numbers are in general use except for the very small sizes. The second column gives the diameter in mils corresponding to each gage number. The third column lists the cross-sectional area in circular mils, rounded off from the square of the diameter in mils according to rules listed below Table A. In the fourth column the resistances per 1000 feet of each size are stated, calculated as outlined above.

Example 8. Use the wire table to find the resistance of 4000 feet of copper wire 0.144 inch in diameter.

Inspection of Table A shows the No. 7 wire has a diameter of almost 144 mils and a resistance of 0.498 ohm per 1000 feet.

$$\text{resistance of 4000 feet} = 4 \times 0.498 = 1.922 \text{ ohms}$$

Example 9. What AWG size of copper wire should be used if 1600 feet must not have a resistance of more than 0.4 ohm?

$$\text{resistance per 1000 feet} = 0.4/1.6 = 0.25 \text{ ohm}$$

From Table A, No. 4 wire (having a diameter of 204 mils) has a resistance of 0.248 ohm per 1000 feet. Thus, the choice should be No. 4 wire.

Note: When the computation demands a wire of a size not in the table, always choose the next LARGER size, i.e., a size having a SMALLER resistance than that computed. In the problems that follow, copper should be assumed as the material unless some other material is specified.

Example 10. A load is to receive 4 kilowatts at 230 volts, over a 1-mile line with a line drop that shall not exceed 10 volts. What AWG size wire is required?

$$\text{current in the line} = \frac{\text{watts}}{\text{volts}}$$

$$= \frac{4000}{230} = 17.4 \text{ amperes}$$

resistance of the line:

$$\text{resistance (line)} = \frac{\text{volts (of line drop)}}{\text{amperes (in line)}}$$

$$= \frac{10}{17.4} = 0.574 \text{ ohm}$$

Since a 1-mile line for 230 volts d-c has two wires, each 1 mile long, the line wire needed is 2 miles long.

$$\text{resistance of 2 miles, or 10,560 feet, of wire} = 0.574 \text{ ohm}$$

$$\text{resistance of 1000 feet} = \frac{0.574}{10.56} = 0.0544 \text{ ohm}$$

From Table A, No. 4/0 wire has a resistance of 0.0490 ohm per 1000 feet, and No. 3/0 wire has a resistance of 0.0618 ohm per 1000 feet. For the voltage drop to be less than 10 volts, the choice must be No. 0000 wire with its diameter of 460 mils.

Prob. 58-9. What AWG size is the wire in Prob. 49-9?
Prob. 59-9. What AWG size is the wire in Prob. 51-9?
Prob. 60-9. What AWG size is the wire in Prob. 53-9?
Prob. 61-9. What AWG size is the wire in Prob. 56-9?
Prob. 62-9. What AWG size is a wire having a cross-section of 0.083 square inch?
Prob. 63-9. What size of copper wire (AWG) must be used to supply a current of 30 amperes to a lamp bank 800 feet distant from the source, with a 3-volt line drop?
Prob. 64-9. What is the resistance per mile of No. 10 wire?
Prob. 65-9. Over what distance can 20 amperes be carried in No. 6 wire with a 2-volt line drop?
Prob. 66-9. What AWG wire has a resistance of about 3 ohms per mile?
Prob. 67-9. How many miles of AWG No. 2/0 wire would be needed to make up a resistance of 6 ohms?
Prob. 68-9. In Fig. 79, assume that the lamps are approximately 1000 feet from the generator. What AWG size wire must be used?
Prob. 69-9. How far would the lamps of Fig. 79 be from the generator if No. 12 wire were used for the line?
Prob. 70-9. If B. & S. gage No. 10 wire were used in Prob. 69-9, how long would the line be?
Prob. 71-9. The line wire in Fig. 118 is No. 12, B. & S. gage. What is the distance between the generator and the lamps?
Prob. 72-9. What size wire is used in Fig. 18 if the motor is 550 feet from the generator?
Prob. 73-9. How many feet of AWG No. 00 wire is needed to provide a resistance of 1 ohm?
Prob. 74-9. What size wire, AWG, would be required to carry a current of 50 amperes over a distance of 800 feet with only a 6-volt line drop?
Prob. 75-9. Each lamp in Fig. 163 is a 110-volt 60-watt bulb. If the lamps are 800 feet from the generator, what AWG size wire must be used?
Prob. 76-9. The motor in Fig. 164 supplies 5 horsepower, with an efficiency of 85%. The line wires are No. 6.

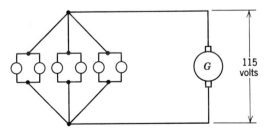

Fig. 163. The voltage at the lamps is lower than the voltage at the generator.

190 Essentials of Electricity

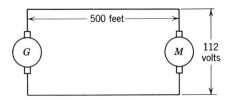

Fig. 164. The difference in voltage between generator and motor is the drop through 1000 feet of wire.

(a) What must be the voltage of the generator?
(b) What capacity must the generator have, expressed in kilowatts?

Prob. 77-9. In Fig. 85, lamp Group I is 300 feet from the generator; lamp Group II is 480 feet from Group I. The generator is operating with a terminal potential of 115 volts. All of the wires are size No. 10. What is the voltage across Group I? What is the voltage across Group II?

Prob. 78-9. If the distances in Prob. 77-9 were twice as great, what size wire could be used for the same voltages across all points?

Prob. 79-9. The resistance of a coil of No. 20 wire, as measured with an ohmmeter, is found to be 15 ohms. How many feet of wire are there in the coil?

Prob. 80-9. A coil for an electromagnet has 800 turns of No. 23 copper wire. The average length of a turn is 6 inches. What is the resistance of the coil?

Prob. 81-9. It is desired to construct a coil of wire that will have not more than 53 ohms of resistance. The coil must have 200 turns of about 16 inches average length per turn. What AWG size wire should be used?

Prob. 82-9. Each lamp in Fig. 165 requires 2 amperes at 112 volts. The rating of the motor is 110 volts, 4 horsepower, at 80% efficiency. What size wire must be used between the motor and lamps?

Prob. 83-9. If No. 4 wire is used between the lamps and generator of Prob. 82-9, what is the voltage of the generator?

Prob. 84-9. What is the total line loss in watts in Prob. 83-9?

Prob. 85-9. How many feet of No. 40 wire are needed for a resistance of 16 ohms?

Prob. 86-9. Calculate the line loss in Prob. 63-9, and express your answer in watts per foot along one conductor.

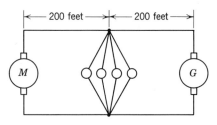

Fig. 165. The voltages at generator, lamps, and motor will all be different.

9. Stranded Wire. Large conductors, especially those larger than No. 4/0, are made of many strands of a smaller wire twisted together. Because of their greater flexibility, such stranded cables are much easier to pull through conduit and are less likely to break when bent at a sharp angle. Special extra-flexible cables are used as train connectors, portable motor leads, and other connections that are subjected to continual motion or vibration.

For example, instead of a solid No. 4 wire with a diameter of 204 mils and an area of 41,700 circular mils, it is sometimes desirable to provide a cable made up of seven wires, each of which is 77.3 mils in diameter and thus 5970 circular mils in area. Since the cable is made up of seven of these strands, the area of the cable is $7 \times 5970 = 41,800$ circular mils, or essentially the area of a No. 4 solid wire. Such an arrangement of wires of equal diameter is known as *concentric stranding*.

For a No. 4 cable of extra flexibility, 49 wires of AWG No. 21 size are arranged seven in a strand, and the seven resulting strands are then twisted together. This arrangement of stranded strands is known as *rope stranding*.

As a consequence of the geometry of concentric stranding of cables, as indicated in Fig. 166, cables are often made up with 7, 19, 37, 61, 91, 127, etc., strands. Successive layers of the smaller conductors are spiraled in alternate directions.

Prob. 87-9. To what AWG size wire is a stranded cable equivalent if the cable has 19 strands, each 0.051 inch in diameter?

Prob. 88-9. It is desired to make a cable of 19 strands which would be equivalent to a No. 2/0 AWG solid conductor. What diameter in mils should each strand have? What AWG size of solid wire should be chosen for each strand?

Prob. 89-9. What is the cross-sectional area in circular mils of the copper in the very flexible 49-conductor cable described in this section? Compare this area to the area of a No. 4 solid conductor.

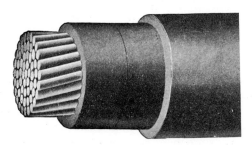

Fig. 166. Count the number of strands in the successive layers of this lead-covered power cable. *American Steel and Wire Co.*

Prob. 90-9. A flexible cable intended for application where vibration is severe has 275 No. 24 copper wires. What size AWG solid conductor is its nearest equivalent?

Prob. 91-9. Specify the number of conductors and their arrangement to produce a cable most nearly equivalent to a No. 6 solid conductor, if the small conductors are: (*a*) No. 23 wires and (*b*) No. 24 wires.

10. Other Conductors. Aluminum is used extensively in the manufacture of conductors for high-voltage power transmission. Although its resistance per circular mil-foot is 17.0 ohms (about 1.64 times that of copper), its density is only 0.3 times that of copper. These facts indicate that an aluminum conductor having the same resistance as that of a given copper conductor will be much lighter than the copper one. For example, a No. 1/0 (105,600 circular mils) solid copper conductor has a resistance of 0.09825 ohm per 1000 feet. An aluminum conductor having the same resistance would have a cross-sectional area of $1.64 \times 105{,}600 = 173{,}200$ circular mils. Its weight would be only $1.64 \times 0.3 = 0.49$ times the weight of the copper conductor, i.e., the equivalent aluminum conductor would weigh only about half as much as the copper conductor.

Example 11. What is the resistance of a No. 6 aluminum wire 2000 feet long?

circular-mil area of No. 6 wire = 26,240 circular mils
resistance of 1 circular mil-foot of aluminum = 17.0 ohms

$$\text{resistance of 1 foot of No. 6 aluminum wire} = \frac{17.0}{26{,}240}$$

$$= 0.000648 \text{ ohm}$$

resistance of 2000 feet of No. 6 aluminum wire:

$$2000 \times 0.000648 = 1.296 \text{ ohms; or}$$

$$\text{resistance} = \frac{\text{resistance of unit wire} \times \text{length}}{\text{circular-mil area}}$$

$$= \frac{17.0 \times 2000}{26{,}240} = 1.296 \text{ ohms}$$

The most popular type of aluminum conductor has one or more steel strands at its center to give it additional tensile strength. This combination of lightness and high tensile strength provides an excellent conductor for bridging long spans such as those encountered in river crossings or in mountainous regions. A cable having this composite construction is called A.C.S.R., meaning aluminum cable steel reinforced.

A steel core inside a copper coating affords a strong solid conductor for carrying small currents. Telephone lead-in wires from pole to house are one example of such steel-cored wire.

For the construction of resistors of large power-handling capacity, iron and steel sometimes find application, but these materials suffer oxidation and become brittle. For heating elements, special high-resistance alloys have been developed with resistances ranging from about 100 to 800 ohms per circular mil foot. A list of the resistances of unit wires constructed of various commercial materials appears as Table B in the Appendix.

For the construction of resistors to be employed in precision measurements, it is important that the resistance value be as independent of temperature as possible. Some of the special alloys listed in Table B were developed to provide this property. For the wire known as Manganin in the temperature range where precision measurements are normally made, the standard material will change its resistance a maximum amount of 15 parts per million, or 0.000015 of the total resistance. Special treatment of this and other alloys can reduce the figure to an even lower value.

Prob. 92-9. What resistance will 1 mile of No. 2 solid copper wire have?

Prob. 93-9. If the wire of Prob. 92-9 were made of aluminum, what would its resistance be?

Prob. 94-9. If the wire of Prob. 92-9 were made of iron, with a resistivity of 75 ohms per circular mil-foot, what would its resistance be?

Prob. 95-9. A No. 4/0 all-aluminum cable is composed of seven strands of 174-mil wires. What is its resistance per 1000 feet, and to what size solid copper conductor is it most nearly equivalent?

Prob. 96-9. A No. 4/0 A.C.S.R. cable composed of six 188-mil aluminum conductors spiraled around a 188-mil steel core carries an alternating current. At the commercial power frequency for alternating current, nearly all of the current is found to crowd outward to lie in the aluminum conductors only. Therefore, neglecting the steel core, what is the resistance of this cable per 1000 feet, and to what size of solid copper conductor is it most nearly equivalent?

Prob. 97-9. A rheostat is designed to have a resistance of 5 ohms. How many feet of No. 24 Driver-Harris Nichrome wire will be needed?

Prob. 98-9. What resistance would a 25-foot length of No. 30 Manganin wire have?

Prob. 99-9. Find the resistance per 1000 feet of No. 23 Driver-Harris Chromax wire.

11. Safe Current-Carrying Capacity for Copper Wires. Since a copper wire has some resistance, a current in the wire will produce a voltage drop across the wire and develop a power loss in the wire. Although the voltage drop may not seriously affect the behavior of the remaining parts of the circuit in which the copper wire is used, the lost power (in generating heat in the wire) may increase the temperature of the wire to a dangerous value. When the temperature of the wire increases above a safe value, the insulation first melts or smolders and perhaps flashes into flame. Further increase in temperature will ultimately reach the melting point of copper, with rupture of the circuit connection when the copper fuses, i.e., softens into a liquid.

If a copper wire is located in a stream of cooling air, it can carry a large current before damage occurs to its insulation. However, if a copper wire of the same size is one of many lying inside a tightly closed conduit located in a very hot area of a factory, the current that will produce a damaging temperature rise is much smaller. Because there are many such complicating factors in the choice of a safe wire size, certain standard recommendations have been developed for normally safe practice. Installations of building wire are governed by the National Electrical Code (NEC), whereas installations of cable for large power-distribution circuits may be arranged in accordance with the recommendations of the Insulated Power Cable Engineers Association (IPCEA). A few typical recommendations are presented in Table C of the Appendix.

Example 12. It is desired to install in a raceway in a building a pair of copper wires suitable for safely carrying 50 amperes in accordance with the specifications of the NEC. What size of copper wire should be used?

From Table C, No. 6 rubber-insulated wire has a capacity of 55 amperes and should be chosen. If regulations permit varnished-cloth insulation, No. 8 wire could be selected.

Prob. 100-9. What size rubber-covered wire should be chosen to carry 15 amperes?

Prob. 101-9. For a conductor temperature of 194°F, what maximum value of current is given in Table C for No. 4 wire?

Prob. 102-9. What is the voltage drop in 200 feet of the wire in Prob. 100-9, when it is carrying 15 amperes?

Prob. 103-9. What maximum current is recommended for a No. 18 rubber-insulated rubber-sheathed portable hard-service cord?

Prob. 104-9. If a portable cord of the rubber-insulated rubber-sheathed type is to be chosen for use with the air conditioner of Prob. 88-7, what size wire is needed to satisfy the recommendations of Table C?

12. Voltage Level and Line Loss. In the planning of a distribution system for electric power, the amount of power lost in the line wires may lead to high costs. If the voltage at which the power is distributed can be increased, the line losses at the higher voltage level can often be reduced.

Power is to be supplied to eight 120-volt lamps, as shown in Fig. 167, each of which requires 0.5 ampere. The length of each line wire is 2000 feet; with No. 7 wire, the resistance is about 1 ohm per line wire. For the parallel connection shown in Fig. 167, the line current is $8 \times 0.5 = 4$ amperes. Because the total line resistance is 2 ohms, the line drop required is $2 \times 4 = 8$ volts. Therefore, the line loss required to supply

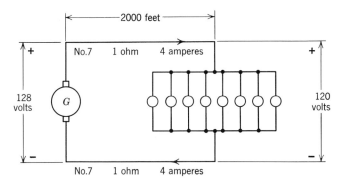

Fig. 167. Each line wire of this two-wire 120-volt system carries eight times the current of one lamp.

the eight lamps in this parallel connection is $8 \times 4 = 32$ watts. For the lamps, the useful power taken is $120 \times 4 = 480$ watts. Thus, to provide 480 watts to the eight lamps of Fig. 167, a line loss of 32 watts occurs as wasted power.

As an alternative connection, consider the arrangement of the eight lamps in Fig. 168, with four paralleled groups each having two lamps in series. The load voltage of the system must now be 240 volts, but the power taken by the lighting load is still 480 watts. However, the line current is now only $4 \times 0.5 = 2$ amperes. Furthermore, the line drop is also cut in half: 2 amperes $\times$ 2 ohms $= 4$ volts, compared with

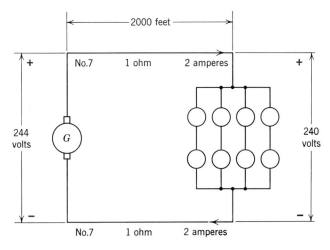

Fig. 168. Each line wire of this two-wire 240-volt system carries four times the current of one lamp.

196   Essentials of Electricity

a line drop of 8 volts for Fig. 167. The line loss for the 240-volt grouping is then only 4 volts $\times$ 2 amperes, or 8 watts. Thus, the 240-volt connection transmits the same 480-watt power to the eight lamps as does the 120-volt connection, but the line loss is only 8 watts for Fig. 168 (compared with 32 watts for the 120-volt transmission).

Note that doubling the voltage makes the line loss one-quarter as great. Thus, if electric power is distributed at 240 volts, the advantage gained through the great reduction in line loss can be employed in *two* ways:

(1) If the same size of line wires is used as would have been required for the 120-volt transmission, the operating costs of the system will be reduced because the line losses are only one-quarter as great. The first cost of installing the equipment will be the same for either connection.

(2) The first cost of setting up the distribution system can be reduced for the 240-volt connection if smaller line wires are chosen. For the same line loss as in the 120-volt system, each line wire could have only one-quarter of the cross-sectional area of the wires originally chosen for the 120-volt system. For the 240-volt connection of Fig. 168, line wires of AWG No. 13 size might have been chosen instead of No. 7. The No. 13 wires would then each have a resistance of about 4 ohms, or a total line resistance of 8 ohms. With a line current of 2 amperes, the line drop would be $2 \times 8 = 16$ volts. The electric power going into heat in the No. 13 line wires would be 16 volts $\times$ 2 amperes $= 32$ watts.

It is clear, then, that transmission of electric power at 240 volts allows a great saving either in the operating cost for line loss or in the original cost of installation over the comparable costs for a 120-volt system transmitting the same power. For the circuit of Fig. 168, the choice of the No. 13 wire instead of No. 7 results in a cross-sectional area of copper one-fourth as large as in Fig. 167, and therefore in a weight of copper only one-fourth as great, with approximately one-fourth the cost of the No. 7 wire.

Similar considerations apply to the distribution of electric power in automobiles. With 6-volt systems, experience has shown that AWG No. 1 cable is usually adequate for normal cranking duty, with No. 1/0 chosen for more severe demands from the battery. For 12-volt batteries, the cables are either No. 4 or No. 6 AWG. Note that if a No. 0 cable were chosen for the 6-volt system, the line drop in the 12-volt system would be about the same if the 12-volt battery were connected with No. 6 cable. The choice of No. 4 cable for the 12-volt system represents a compromise; the cable cost is not as low as it might be, but the line loss between the battery and the starting motor is smaller than was tolerated in the 6-volt system. For the automotive system, of course, the cost of the line loss is not important, but the line drop is of the utmost importance, perhaps making the difference between whether or not the engine of the auto can be started in cold weather.

13. Three-Wire System. Since doubling the voltage results in a great reduction in line loss (or a comparable saving in the initial cost of copper if the wire size is decreased), there is ample justification for making electric power available at 240 volts. Although d-c distribution is no longer very widespread, almost similar considerations apply to the distribution of commercial a-c power. Thus, whereas the a-c circuit is more important, the d-c circuit at 240 volts will be studied here as an introduction to the basic ideas that are involved in the commercial a-c circuit.

For lighting circuits, standardization is essentially so rigid that lamps are commonly designed for 120-volt operation. If this voltage is doubled to take advantage of possible savings, the 240-volt line potential requires that two 120-volt lamps must always be connected in series. Such a connection would require that all lighting fixtures have bulbs arranged to be switched off and on in multiples of two. To avoid this difficulty while retaining the advantage of the 240-volt distribution scheme, a third wire known as the neutral wire is added to the basic two-wire circuit.

The lamps shown in Fig. 168 may be rearranged as shown in Fig. 169 to provide the neutral connection. The neutral wire is often of the same size as each of the line wires. When the lighting load on each side of the neutral is the same, the neutral carries no current. Under these conditions, if the size of each line wire is chosen to yield the same total line loss as in Fig. 168, the circuit of Fig. 169 will have just three-

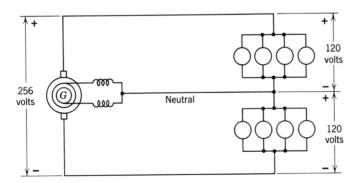

Fig. 169. Each outside wire of this balanced three-wire 240-volt system carries four times the current of one lamp, and the neutral wire carries no current.

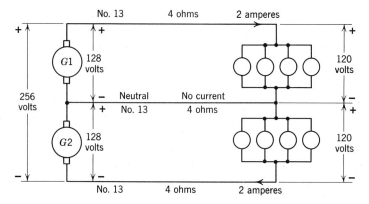

Fig. 170. An equivalent-generator scheme for representing the three-wire generator of Fig. 169.

eighths as much copper as the corresponding 120-volt system that supplies the same total number of lamps.

In many practical three-wire d-c installations, the d-c generator must have a special device to which the third wire is attached, as indicated in Fig. 169. However, for calculations relating to the line wires and the lighting circuits, it is sufficient to consider that the 240-volt generator has been divided into two smaller generators, as shown in Fig. 170, where each generator has a nominal 120-volt designation. The neutral wire is then joined to the common junction between these two fictitious sources.

14. Unbalanced Three-Wire System. As noted above for Fig. 169, the neutral carries no current when the number of lights turned on is the same on each side of the neutral. When a three-wire system has this equality of loading, the system is said to be *balanced*. If the loads are not equal, the neutral wire does carry current and the system is said to be *unbalanced*. Thus, the neutral becomes useful in an unbalanced system, i.e., when the appliances connected on one side of the neutral are carrying more current than those on the other side, with the surplus appearing in the neutral. For the unbalanced loading shown in Fig. 171, four 0.5-ampere lamps form the load on the + side of the neutral, and only one 0.5-ampere lamp is the load on the − side. Thus, the neutral in Fig. 171 must return a current of 1.5 amperes to the source junction, and generator $G1$ supplies the greater part of the total lighting power. In contrast, as in Fig. 174, if the four lamps are lighted on the − side of the neutral, and if there is only one lamp lighted on the + side, the

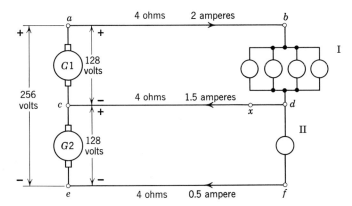

Fig. 171. With unbalanced loading, the neutral wire of the three-wire system carries some current.

neutral carries a current of 1.5 amperes from the source junction to the lighting load, with generator G2 supplying most of the power.

15. Voltage Diagrams for Unbalance. In operating three-wire systems that distribute large blocks of power, every effort is made to keep the loading balanced. For such a balanced system, the values of voltage, current, and power would differ in no respect from those of a corresponding two-wire system with all loads paired off in series within many paralleled groups, as Fig. 167 indicates for a lighting load. On a large system, the connecting or disconnecting of a few lights makes no noticeable difference, but it is instructive to see how the voltages of a three-wire system change for gross unbalance like that of Fig. 171. Such changes can be most readily visualized with the construction of voltage diagrams in the manner discussed in Section 6-4.

Consider the lamps of Fig. 171 to be constant-current loads of 0.5 ampere each. The drop in the upper 4-ohm line wire from point *a* to point *b* is 2 amperes × 4 ohms = 8 volts. The drop in the neutral from point *d* to point *c* is 1.5 amperes × 4 ohms = 6 volts. Applying Kirchhoff's voltage law around the upper mesh of this circuit (*abdca*) yields the drop from point *b* to point *d*, or the load voltage: 128 − 8 − 6 = 114 volts. Figure 172 shows the voltage diagram for this upper mesh of the distribution circuit. Note that the line drop is 8 + 6 = 14 volts, and it is *not* equally divided between the connecting wires as in the more simple two-wire systems.

In the lower mesh of the circuit (*cdfec*), the drop from point *f* to point *e* is 0.5 ampere × 4 ohms = 2 volts. For this mesh, Kirchhoff's voltage

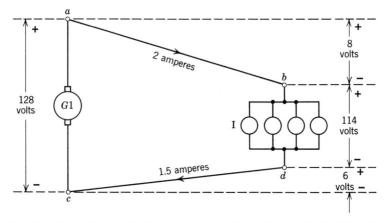

Fig. 172. Voltage diagram for the upper mesh of the unbalanced circuit of Fig. 171.

law must be applied with care. The sum of the *rises* in voltage from point e to point d is 128 + 6 = 134 volts. The sum of the *drops* from point d to point e must be equal to this value, or (voltage *df*) + (voltage *fe*) = 134 volts. Therefore, the load voltage (from d to f) must be 134 − 2 = 132 volts. The voltage diagram of Fig. 173 shows these relationships clearly. Note that the line "drop" is now a rise of 4 volts for this lower mesh, because the load voltage is 4 volts higher than the source voltage.

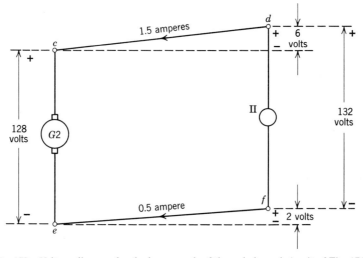

Fig. 173. Voltage diagram for the lower mesh of the unbalanced circuit of Fig. 171.

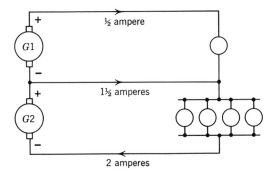

Fig. 174. Unbalanced loading on a three-wire system requires current in the neutral wire.

Thus, gross unbalance of a three-wire system tends to raise the voltage of the load requiring less power to values larger than the value of the half-generator voltage. Whereas the increase shown here is only 10% of the original 120-volt potential that was applied to each lamp in the balanced system of Fig. 170, some lamps and certain electronic tubes would have their useful life considerably shortened by continued application of such an overvoltage. Thus, the result of extreme unbalancing of a three-wire system is a change in the load voltages sufficiently large that appliances designed for special standard voltages may not operate satisfactorily with such unbalance.

Prob. 105-9. Each of the three wires in Fig. 174 has a resistance of 2.5 ohms. Each of the two generators maintains a constant terminal potential of 125 volts, and each lamp requires 0.5 ampere. Calculate all voltages and construct the voltage diagrams for the upper and lower meshes for this circuit.

Prob. 106-9. In Fig. 171, assume that the loads are changed, but that the voltage of each generator and the resistance of each line wire remain the same. For Group I there are to be six lamps, and for Group II only two lamps, with each lamp requiring 0.5 ampere. Calculate the voltage across each load group.

16. Open Neutral. A possible danger in the operation of a three-wire system is the chance that the neutral may become broken at the same time that the load is unbalanced. Although such rupture is not at all common, the resulting voltage changes are worth studying for an understanding of what does happen when such an accident occurs.

In Fig. 171 the point (x) marks a location where the neutral might be broken. A break in the neutral wire at point (x) leads to the series circuit of Fig. 175. In the calculations on this circuit, the resistance of each lamp will be assumed to remain constant at 240 ohms. For the four lamps in parallel in Group I, the combined resistance is $\frac{240}{4} = 60$

202 Essentials of Electricity

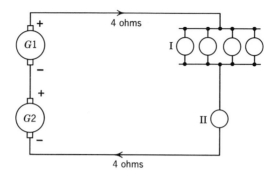

Fig. 175. When the neutral wire is accidentally opened, a series circuit results.

ohms. Since Group II has only one lamp, the load resistance of Group II is 240 ohms. The series connection of these two load groups leads to a total resistance of the loads of 60 + 240 = 300 ohms.

The two remaining line wires still have a total resistance of 8 ohms. When this value is added to the load resistance, the total resistance connected across the generator terminals a and e is 8 + 300 = 308 ohms. As in Fig. 171, the total voltage between points a and e is 256 volts, leading to a line current of $\frac{256}{308}$ = 0.831 ampere.

Since the line current is also the load current in this series connection, Ohm's law applied to each lamp group will yield the separate load voltages. Across Group I, the voltage is 0.831 × 60 = 49.9 volts. For Group II, the load voltage is 0.831 × 240 = 199.4 volts. The low voltage across Group I where the original power consumed was large will now hardly cause the lamps to glow. On the other hand, the potential of almost 200 volts applied across the single 120-volt lamp in Group II will shortly burn out the lamp and disconnect the entire circuit. This calculation demonstrates clearly the importance of maintaining a working connection of the neutral wire of a three-wire system if the loads on each side of the neutral can become unbalanced.

Prob. 107-9. Calculate the voltage that would appear across each load group in Prob. 105-9 if the neutral wire were opened.

Prob. 108-9. For a broken neutral in the circuit of Prob. 106-9, calculate the new load voltages.

Prob. 109-9. Calculate the load voltages in Fig. 170 with the neutral opened.

SUMMARY

Wire is usually made of COPPER or ALUMINUM because of their low resistance.

If the resistance of 1 foot of any size wire is known, the resistance of any length is found by MULTIPLYING BY THE LENGTH in feet.

9: Wire and Wiring Systems

If the resistance of 1 foot of wire $\frac{1}{1000}$ inch in diameter is known, the resistance of 1 foot of wire of any diameter can be found by DIVIDING BY THE SQUARE OF THE DIAMETER in thousandths of an inch.

A MIL is $\frac{1}{1000}$ inch.

A CIRCULAR MIL is the area of a circle 1 mil in diameter.

The area of any circle in circular mils equals the SQUARE OF THE DIAMETER in mils.

A UNIT WIRE is called a circular mil-foot; it is a round wire 1 foot long and 1 mil in diameter.

The resistance of a circular mil-foot of copper wire is 10.4 ohms.

The resistance of a round copper wire equals

$$\frac{10.4 \text{ (resistance of a circular mil-foot)} \times \text{length in feet}}{\text{circular-mil area}}$$

The resistance of standard sizes of copper wire may be found from WIRE TABLES.

Wire is often STRANDED for greater flexibility. The gage number of stranded wire depends upon the total circular-mil area of all the strands, the stranded cable being essentially equivalent to a solid wire of the same total area.

The resistance of a circular mil-foot of aluminum wire is 17.0 ohms.

The resistance of a circular mil-foot of special alloys may be as high as 800 ohms.

For interior wiring, insuring agreements require that the choice of the wire size must be determined by the National Electrical Code. The heating effect of current imposes a limit for safe operation of a given wire size.

The same amount of power can be transmitted over the same wires at one-fourth the line loss in watts if a potential of 240 volts is used instead of 120 volts.

The same amount of power can be transmitted at the same loss in watts over wires having only one-fourth the weight (or area) if a line potential of 240 volts is used instead of 120 volts.

The THREE-WIRE SYSTEM permits taking advantage of either of the savings noted above, without preventing the use of 120-volt appliances.

The NEUTRAL carries current only when one line wire carries more current than the other. Such an UNBALANCED CIRCUIT results in an uneven distribution of voltage to the load groups. A broken neutral with an unbalanced load may ruin appliances attached to the line.

PROBLEMS

Prob. 110-9. How many 120-volt lamps, each requiring 0.5 ampere, can be connected to a circuit in which the power taken may not exceed 1320 watts?

Prob. 111-9. What size wire must be used for the circuit of Prob. 110-9?

Prob. 112-9. What will be the line drop in No. 14 wire for the loading of Prob. 110-9 if the distance between the distribution point and the lamp group is 300 feet?

Prob. 113-9. Seven No. 31 copper wires are stranded into a cable. To what AWG size is the cable equivalent?

Prob. 114-9. How many strands of No. 10 wire should there be in a cable that is equivalent to a No. 6 solid wire?

Prob. 115-9. How many strands of No. 32 wire will it take to make a cable equivalent to a No. 5 solid wire?

Prob. 116-9. What is the diameter of a solid wire having an area of 250,000 circular mils?

Prob. 117-9. How many No. 18 copper strands will it take to make a cable equivalent to the solid wire of Prob. 116-9?

Prob. 118-9. What is the safe current-carrying capacity of the cable of Problem 117-9, as specified in the National Electrical Code?

Prob. 119-9. What is the safe current-carrying rating from the National Electrical Code for the cable of Prob. 113-9?

Prob. 120-9. What AWG size of aluminum wire is equivalent in resistance to No. 10 copper wire?

Prob. 121-9. What is the safe current-carrying rating of the aluminum wire in Prob. 120-9 if rubber covered?

Prob. 122-9. Consider that the current in each lamp in Fig. 176 is constant at 1.5 amperes. Each of the two generators has a constant terminal potential of 112 volts. Find: (a) the line drop from $G1$ to Group A, from Group A to Group B, and from $G2$ to Group C, (b) the voltage across each set of lamps, and (c) the power delivered by each generator.

Prob. 123-9. Assume that the resistance of each lamp in Fig. 176 is constant at 60 ohms. Each generator maintains a constant terminal potential of 112 volts. If the neutral becomes open between points o and s, find: (a) the voltage across each set of lamps, (b) the power delivered by each generator.

Prob. 124-9. If the break in the neutral occurs between points s and v in Prob. 123-9, find the new values for parts (a) and (b).

Prob. 125-9. At a point 2 miles from a generator the voltage is to be 550 volts

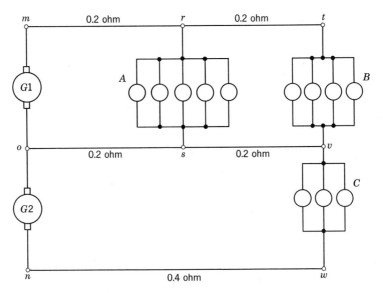

Fig. 176. Unbalanced loading on a three-wire system having an additional load (A).

9: Wire and Wiring Systems

when a load is consuming 150 horsepower. The watts lost in the line are not to exceed 7% of the watts delivered to the load. What diameter of solid copper wire would be required?

Prob. 126-9. What diameter of aluminum wire would be required in Prob. 125-9?

Prob. 127-9. What size of rubber-covered aluminum conductor is required for interior wiring that carries 35 amperes?

Prob. 128-9. A bank of 200 lamps, each of which is rated at 100 watts at 110 volts, is to be supplied from a source having a terminal potential of 118 volts. If the lamps are 800 feet from the source terminals, what size copper wire is required for each of the two line wires?

Prob. 129-9. Hard-drawn copper has a resistance of 10.7 ohms for a circular mil-foot wire at 68°F. When stranded, its resistance is further increased by approximately 2%. What is the resistance of 1000 feet of a hard-drawn copper cable comprising 37 strands of 116.2-mil wire?

Prob. 130-9. What is the resistance per 1000 feet of an all-aluminum cable made up like the copper cable of Prob. 129-9?

Prob. 131-9. What is the resistance per 1000 feet of an all-aluminum cable comprising nineteen 162.3-mil conductors?

Prob. 132-9. On very hot days the temperature of aerial transmission-line conductors runs over 100°F. When the temperature is 122°F., the resistance of hard-drawn copper is 12.2 ohms for a circular mil-foot. Find the change in resistance when the temperature of the cable of Prob. 129-9 changes from 68° F. to 122° F.

Prob. 133-9. Find the rise in resistance of the cable of Prob. 130-9 as the temperature rises from 68° F. to 122° F. if the resistance of commercial aluminum at 122° F. is 19.3 ohms per circular mil-foot. What percentage of change is this amount?

Prob. 134-9. For an A.C.S.R. cable, the resistance of the steel core may be neglected in calculating the cable resistance. Find the resistance of 1 mile of 500,000 circular-mil A.C.S.R. cable composed of thirty 129.1-mil aluminum strands and seven steel strands.

Prob. 135-9. How many feet of No. 18 Driver-Harris Chromax wire will be required to make a 10-ohm resistor?

10
Switches and Signal Devices

1. Switches, Relays, and Contactors. Electric devices are especially useful if they can be controlled remotely or automatically. The simplest device for connecting and disconnecting an ordinary household circuit is the toggle snap switch, of the kind shown in Fig. 177. This device provides a movable contact for one side of a two-wire circuit, the other wire often being grounded. Since only one side of the circuit is switched, this form of switch is called a *single-pole* switch. A schematic representation for such a switch was given in Fig. 1, as well as the symbol for another type of switch (*double-pole*) that permits opening both wires of the circuit simultaneously.

For certain types of automatic or remote switching, it is important to have a small current in one circuit actuate a device that will close switch contacts for a much larger current in a separate circuit. When the power level of the controlled circuit is small, the switching device is usually known as a *relay*. A magnetic relay having several *contacts* in its controlled circuits is shown in Fig. 178; with such relays many switching operations can be controlled by a single separate current. The coil of the relay sets up a magnetic field in the iron yoke of the device, and the force of magnetic attraction pulls the moving part of the yoke closed against the fixed part of the iron. An insulated member attached to the moving part of the yoke pushes the controlled contacts out of their normal arrangement, closing some circuits and opening others. When

10: Switches and Signal Devices

Fig. 177. Single-pole toggle switch: (a) OFF, (b) ON. *Harvey Hubbell, Inc.*

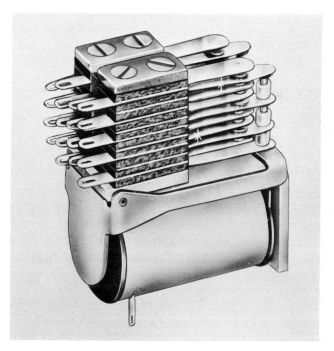

Fig. 178. Miniature relay for business machines and airborne computers. *Potter & Brumfield Div., American Machine & Foundry Co.*

208 Essentials of Electricity

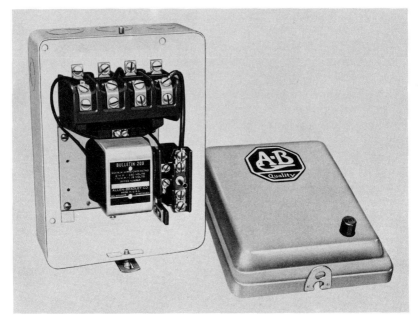

Fig. 179. This contactor for full-voltage motor starting is equipped with an overload device that can be reset by pushbutton. *Allen-Bradley Company.*

the current in the relay coil is cut off, a spring or the pull of gravity draws the moving elements back into normal position again.

If the power level of the controlled circuit of a relay is large, the relay is then usually known as a *contactor*, especially if it must operate frequently in switching the large power. Figure 179 shows a form of contactor employed for connecting motors directly to line wires in applications where this operation is possible.

2. Basic Contact Connections. Although designations of single-pole, double-pole, and triple-pole are commonly employed to describe simple switches, such a system becomes cumbersome when there are many wires to be switched. In systems involving many connections, attention is focused on the *number* of contacts; in turn, each contact is given a special designation to make its operation clear. When all controlling power is removed from a relay or other switching device, and when any associated springs or latches have been let down, the device is said to be in its *normal* condition. The nature of the contacts in the controlled circuits is then described either as *normally open* or as *normally closed*. It is important to fix this definition clearly in mind. For example, the contactor of Fig. 179 may be arranged to keep a motor running for 23 hours, so that the motor is only OFF for 1 hour each day. Nevertheless, this contactor is described as having normally open contacts, in accordance with the definition stated above. Figure 180 shows how normally open (*NO*) and normally closed (*NC*) contacts are represented in circuit diagrams.

A common motor-control device that usually has one of each kind of contact is the *pushbutton station* shown in Fig. 181. In the usual circuit employing this device, the START button operates a normally open contact. When the START button is released, a spring holds the

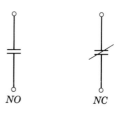

Fig. 180. Circuit representation of basic contact connections.

Fig. 181. Heavy-duty pushbutton station. *Allen-Bradley Company.*

contact pair open. Depressing the button causes the normally open contact to be closed as long as the button is held down. There is usually a normally closed contact under the control of the STOP button. Depressing the STOP button displaces the contact from its normal condition, thus opening the circuit.

Of course the push buttons are arranged so that they need to be depressed only temporarily to bring about the desired automatic operation. One method of leaving a controlled circuit connected after a START button has been depressed and then released involves a *latching relay,* or *latching contactor.* The coil of the latching relay is placed in series with the normally open START button and a source of electric power. Depressing the START button causes the relay to operate and start the system connected in the separate circuits of its contacts. However, a mechanical latch holds the moving member of the relay in the operated condition, leaving the controlled circuits functioning when the START button is released. With this arrangement, the STOP button must be connected in such a way that an auxiliary device unhooks the latch when the equipment is to be shut down. Although this complexity is a disadvantage of this scheme, the latching relay requires no current in its coil except when it is operated. Such an arrangement may justify the complication for the motor running 23 hours a day, as described above.

An alternate method for arranging the circuit of the momentary push-button station employs a separate normally open contact on a nonlatching relay. In Fig. 182, the circle containing letter M represents the coil of the main contactor of a motor-starting arrangement. The heavy-current contacts are in a separate circuit not shown. However, the auxiliary normally open contact labeled M operates at the same time as the motor-starting contacts. When the circuit is completely de-energized, contactor M is in its normal condition, with no current

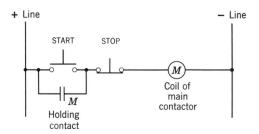

Fig. 182. The holding contact permits operating the contactor with only a brief touch on the START button.

through its operating coil. When the START button is depressed, its normally open contact is temporarily closed, providing a circuit to set up current through the normally closed STOP button into the coil of contactor M. Then contactor M operates to start the motor (not shown), and also to close the auxiliary *holding contact M*. Thus, the contactor holds or seals itself onto the power source, allowing the START button to be released, while the motor circuit continues to operate. As long as the controlled motor circuit is to remain on, there must be current in the coil of contactor M supplied through the holding contact M and the normally closed contact of the STOP button. When the motor circuit is to be shut down, depressing the STOP button will interrupt the current to the coil of contactor M. If the STOP button is held down long enough for the moving part of the contactor to open the normally open holding contact, the circuit will become completely de-energized again. Furthermore, this arrangement of the push buttons avoids any accident if both buttons are depressed simultaneously. When the circuit is de-energized, depressing both buttons leaves the circuit de-energized. If both buttons are depressed while the motor is running, the circuit will be de-energized exactly as if only the STOP button had been pushed. Practical magnetic controllers for motor circuits also have other contacts included in series with the STOP button to cause shutdown in case of danger to the motor due to overloading.

3. Electric Bells. Many signaling devices employ electric bells to give notice of some remote action in the system. The simplest *single-stroke bell* is constructed as shown in Fig. 183. When button P is pushed, the battery establishes a current through the coils of electromagnet M. Flux is set up in the iron yoke, with lines directed to the right through the lower coil, up through armature K, and back to the left through the upper coil. Recall from Chapter 4 that Rule A stated that magnetic flux lines form closed loops. These closed loops represent magnetic forces attempting to draw unlike poles together. Thus, the magnetic attrac-

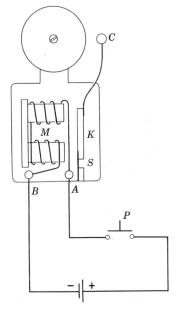

Fig. 183. Single-stroke electric bell.

tion draws armature K over toward the magnet, causing clapper C to strike the bell. As long as button P is kept depressed, armature K remains against the magnet. When the button is released, its contact returns to its normally open position. The circuit through the electromagnet is opened, and spring S returns armature K to its initial position. This type of bell produces only a single stroke each time button P is depressed. Single-stroke bells are often used when coded ringing must be sounded in many parts of a building. In such applications a normally open contact of a master contactor may replace button P. Electric chimes are also operated by single-stroke mechanisms, although usually of a more compact design.

The arrangement of elements for a vibrating-stroke bell is shown in Fig. 184. Armature K now carries one element of a normally closed contact pair, the other element being fixed by set screw G. When button P is depressed, current is set up by the battery through the normally closed contact at G to the electromagnet. The resulting magnetic attraction draws armature K over to the left, causing clapper C to hit the bell. However, this motion of armature K opens the normally closed contact at G, thereby de-energizing the electromagnet. Spring S returns armature K to its normal position. With button P still depressed, the contact at G closes again, and the action repeats. Armature K continues to vibrate rapidly, with clapper C striking a succession of blows on the gong, until button P is released. Vibrating-stroke bells are generally inexpensive, often operating from relatively low-voltage sources. Although dry cells are a possible source, several schemes permitting connection to a low-voltage a-c source are more common.

Sometimes vibrating-stroke bells are constructed with a latching holding contact. This normally closed contact is latched into the open condition by depressing a reset button on the bell. The latch is hooked to the armature of the bell. As long as the controlling push button is not operated, the bell remains silent. Depressing the push button causes the bell to ring. The motion of the armature of the bell unlatches the holding contact, which

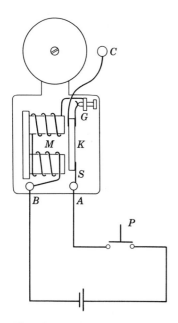

Fig. 184. Basic vibrating-stroke electric bell.

10: Switches and Signal Devices 213

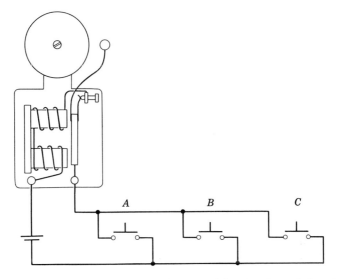

Fig. 185. A bell arranged to be rung from any of three locations: A, B, or C.

then becomes closed. Since this holding contact is connected in parallel with the controlling push button, the bell will continue to ring after the push button is released. Such a bell is useful in an alarm system in which only a momentary contact of the controlling button is possible. Ringing continues until someone depresses the reset button on the bell itself, thereby setting the latch to keep the holding contact open once more.

Example 1. How can a circuit be arranged so that a bell can be rung from any of three locations?

The bell in Fig. 185 can be rung from any of the stations marked A, B, or C. All the push-button stations are connected in parallel.

Prob. 1-10. A bell has three terminals so that it can be used either as a single-stroke or as a vibrating-stroke bell. Draw a diagram showing the wiring of the bell, and its connection to a battery and two push buttons to cause either operation from one location.

Prob. 2-10. Draw a complete diagram of the latching arrangement for the continuous-ringing bell described in this section. Show the bell wiring, the reset button, a battery, and the operating push button. Label the necessary springs and latches clearly.

4. Buzzers. A buzzer is usually constructed like a vibrating stroke bell, as indicated schematically in Fig. 42 (of Chapter 4). The clapper and bell are left off, and the noise is made by a vibrating armature. Sometimes a master buzzer is used to operate a set of inexpensive bells.

Each of the bells is the single-stroke type, and they are all connected in series with each other and with the power source. The buzzer merely opens and closes the circuit, causing the bells to act as vibrating-stroke bells. Maintenance is very simple for such a system because only the contact points on the buzzer need to be kept in good condition.

Prob. 3-10. Draw the diagram of a buzzer operating three bells from a battery. Include the complete electrical circuit.

Prob. 4-10. Two or more circuit-breaking bells will not work well in series. Draw a diagram showing how a single-stroke bell can be used in series with a circuit-breaking bell.

5. Electric Lock Opener. In apartment houses it is desirable that a button in each apartment can release the lock on the door at the main entrance. One form of device that accomplishes this result easily is the mechanism of the single-stroke bell, without the clapper and bell. When the circuit through the single-stroke mechanism is closed by pushing a button in any of the apartments, the movement of the armature releases a special lock and allows the door to be pushed open.

Prob. 5-10. Show the wiring diagram of an arrangement for ringing a bell in an apartment by a push button located at the main-entrance door, together with the diagram for operating the entrance-door opener from a push button in the apartment. Both circuits are to be supplied from the same battery or other appropriate power source.

Prob. 6-10. Show the connections for wiring four bells and one battery so that one bell may be rung from any one of three stations, while the other three bells may be rung from a fourth station.

Prob. 7-10. Show the diagram for wiring the following arrangement:

   Bell No. 1, to be rung from the front-door push button.
   Bell No. 2, to be rung from the rear-door push button.
   Buzzer, to be rung from the dining-room push button.

There is to be only one power source for the entire arrangement.

Prob. 8-10. Show by a diagram a set of three bells which may be operated by one push button.

6. Annunciators. For the typical call-service found in industry, hotels, restaurants, and hospitals, a single buzzer (or bell) is located at a central point. When a push button at any of the remote points is pressed, the buzzer sounds: it is desirable to have a visual indication to show which remote point was signaling. A device that gives such an indication is known as an *annunciator*.

A wall-mounted annunciator for connection to nine remote stations is shown in Fig. 186. Beneath each number on the panel is a round opening, or window. When the push button at a remote station is pressed, a white target, called a *drop,* appears and remains in view in

10: Switches and Signal Devices 215

Fig. 186. Nine-station flush-mounted annunciator. *S. H. Couch Company, Inc.*

the window, even though the button is depressed only momentarily. Although the buzzer associated with the annunciator sounds only momentarily, the visual indication remains until the reset button, shown at the bottom of the panel, is pressed; all drops that are down return to their location out of sight behind the panel. In more complex installations, flashing lights may be provided for visual indication; a private telephone system may also be incorporated with the annunciator to furnish voice communication as well as visual signaling.

The operating mechanism for each drop of Fig. 186 is shown in Fig. 187. There are two electromagnets (relay coils) mounted side by side, the left coil being wound with larger wire than that of the right coil. The left coil is the *operating coil,* and the right coil is the *reset coil.* When either coil is energized, it establishes magnetic flux that crosses from its pole piece to the movable arm (or *armature*), which is pivoted at a point below the poles on a center line between the poles. Two metal fingers project from each end of the armature toward the drop. Although the drop, bearing the white target, is pivoted at the same point as the armature is, the drop can move separately from the armature.

In Fig. 187 the drop and armature are shown in the reset position, i.e., the target would not show through its window. When the left coil is energized from its remote push button, the armature is drawn up

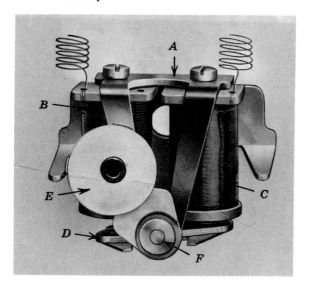

Fig. 187. Annunciator mechanism: *A*, frame; *B*, operating coil; *C*, reset coil; *D*, armature; *E*, target; *F*, drop pivot. *S. H. Couch Company, Inc.*

quickly under the left pole piece. The left finger on the armature accelerates the drop to the right; whereas the armature stops upon striking the left pole piece, the drop continues to move until it stops in place in its window, at rest against the right finger on the armature. Energizing the reset coil reverses this operation, throwing the drop back upward to the left, out of sight behind the panel. Since the reset coils are connected in parallel, their wire size is made small to permit resetting with a small current for each reset coil. With as many as 20 drops per reset push button, this design eliminates the need for a d-c supply of large current capacity. Standard potential for the d-c supply is usually 6 volts.

Figure 188 shows a schematic diagram for the connections for a three-drop annunciator, without reset coils. In some designs, resetting may be accomplished manually with a lever that manipulates all the drops directly into reset position. However, in some installations it may be impossible to reach a manual reset lever. When the annunciator is located high on a wall for easy visibility, the electric reset scheme (with a remote push button) is imperative.

Prob. 9-10. Show a diagram for the internal wiring of a three-drop annunciator with electric reset. Include the wiring for a remote reset push button.

Prob. 10-10. With the annunciator of Fig. 188, a buzzer is to be located at each of the three remote push buttons. Three other push buttons are to be lo-

10: Switches and Signal Devices 217

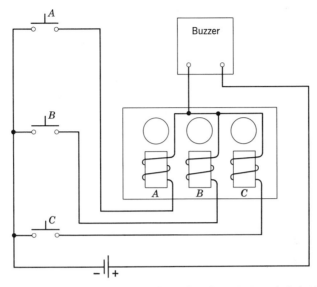

Fig. 188. Annunciator that indicates by which of stations A, B, and C the buzzer has been rung.

cated at the annunciator. This arrangement is known as a return-call annunciator system; for example, a resident in an apartment may signal the superintendent at the annunciator, and the superintendent may acknowledge (or "return") the call by pressing the appropriate button at the annunciator. Show the complete wiring diagram, with the circuit arranged to operate from only one d-c source or battery.

7. Control of Incandescent Lamps. Figure 189 shows how a lamp is wired to a single wall-mounted snap switch, which controls the lamp from only one location. The National Electrical Code requires that grounded conductors shall not ordinarily be disconnected. Therefore, only a single-pole single-throw snap switch (a switch having one normally open contact when in the OFF position) is required, and it must be connected in series with the ungrounded wire of the two-wire connection.

Sometimes it is desirable to be able to operate a lamp from two locations, with the operation at each location being independent of the

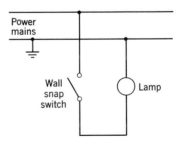

Fig. 189. Lamp controlled from a single wall switch.

position of the switch at the other location. Such a requirement may be met with the use of snap switches known as *three-way* switches. Although the mechanical construction may differ from one form of three-way switch to another, the essential switching element may be designated as a single-pole double-throw element. Usually, such three-way switches have no ON-OFF markings, for this aspect of their operation depends on the position of the three-way switch at the other location in the system. However, if one of the positions of each of the two three-way switches is designated as "normal," it would be possible to say that a three-way switch had one normally open contact and one normally closed contact. Figure 190 shows how two three-way switches, *A* and *B*, are arranged to control a single lamp. The lamp might be used to light a stairway, with one switch upstairs and the other down. Note that the switch points have been arranged to sever connections only to the ungrounded conductor of the system, in accordance with the National Electrical Code.

When a single-pole switch makes contact to several terminal points in succession, the switch is often known as a *rotary switch*. Some lamp fixtures for household use provide switching for a special light, called a *three-way bulb*. Such a bulb provides three different light intensities from one socket. Since the associated switch in such fixtures is known as a *three-way switch,* the two uses of this name must always be distinguished carefully. For lamp fixtures, the name means a rotary switch have three ON positions, and one OFF position. For the control of fixed lighting from two separate locations, the name of *three-way switch* refers to a single-pole double-throw switch, as shown in Fig. 190.

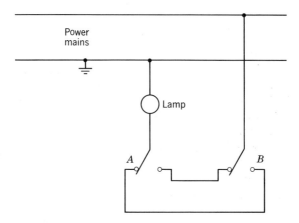

Fig. 190. The lamp may be turned ON or OFF from either switch *A* or switch *B*.

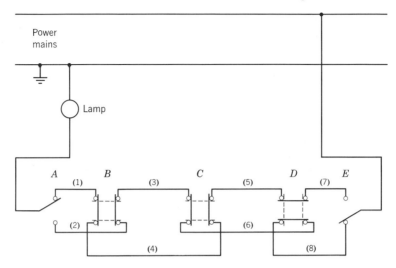

Fig. 191. The lamp may be turned ON or OFF from any of the five locations: A, B, C, D, E.

Figure 191 shows the method of control from any number of locations. Switches A and E are three-way snap switches, and the intervening switches, B, C, and D, in the controlled side of the circuit are called *four-way* switches. The four-way switches have two movable contacts, operated together by the toggle button, to make contact in pairs with the four circuit connections for each switch, as shown in Fig. 191. Note that this connection must similarly not open the grounded conductor to the lamp itself.

Lighting systems in public halls and lecture rooms may require control of the lighting from more than one location. Even when the lamps are controlled from only one central point, the total current required by all the lamps may be too large to be switched successfully by snap switches. For such applications, the snap switch or other single-pole device is connected to the coil of a contactor. The main contacts of the contacting device are then chosen to be large enough to handle the total current satisfactorily. If the contactor is constructed so that all its normally open contacts open simultaneously, the National Electrical Code does permit the grounded conductor to be disconnected at the same time the ungrounded conductor is opened.

For some lighting systems, latching contactors may be desirable to avoid having the contactor hold in for long periods of time. Since power is needed in some other circuit to unlatch the main circuit, such systems usually have an auxiliary power source (often a storage battery) available for unlatching in an emergency. Furthermore, latching contactors will usually have a trip that can be operated by hand in a dire emergency.

Prob. 11-10. Show a diagram of connections for the control of all the lamps in a movie theater. The main switch is to be a magnetic contactor, controlled by two three-way switches, with one of the snap switches in the projection booth, and the other snap switch on the stage.

Prob. 12-10. Show a circuit diagram for controlling the lights in a large auditorium with a contactor and the push-button station of Fig. 181. The buttons would probably be labeled ON in place of START, and OFF in place of STOP. Show a holding contact on the main contactor, and show its wiring in the circuit.

Prob. 13-10. Show a diagram for the control of three lights, one at the top of each of three flights of stairs. The three lights are to go ON simultaneously, and OFF simultaneously. The control shall be from four snap switches, one located at each of the four floors.

Prob. 14-10. Verify that the lamp is OFF in Fig. 191. List, by the numbers shown in parentheses, the wires through which current is supplied to the lamp when switch C is thrown to its opposite position.

SUMMARY

SWITCHES may be classified in terms of the number of parts of the circuit that are switched, i.e., SINGLE-POLE, DOUBLE-POLE, etc.

A RELAY is a device in which a small amount of power controls the switching of a larger amount of power.

CONTACTS are the switching elements in the controlled circuits of a relay.

CONTACTORS are relays designed for switching large quantities of power.

The NORMAL condition of a relay or contactor is its state when all controlling power has been removed and when all associated springs or latches have been let down.

A NORMALLY OPEN CONTACT means a contact-pair that is open in the normal condition of a relay.

A NORMALLY CLOSED CONTACT means a contact-pair that is closed in the normal condition of a relay.

A PUSH BUTTON is a manually operated switch, often having just one normally open (or normally closed) contact.

A HOLDING CONTACT means a contact pair arranged to keep a contactor energized following only momentary energizing through a push button or other similar contact.

A LATCHING CONTACTOR means a contactor that will latch mechanically into its operated condition following only momentary energizing; the latch must be released manually or electrically to return the contactor to its normal condition.

ELECTRIC BELLS have two principal forms: SINGLE-STROKE and VIBRATING-STROKE.

The magnetic mechanism of an electric bell is the essential part of a BUZZER, LOCK-OPENER, and an ANNUNCIATOR.

A THREE-WAY SWITCH is a single-pole double-throw switch.

FOUR-WAY SWITCHES have two movable contacts that make connections by pairs among the four terminals of the device.

Operation of a lamp from TWO or MORE locations is accomplished with circuits connected through three-way and four-way switches.

PROBLEMS

Prob. 15-10. In a certain single-stroke electric bell, the electromagnets have a total of 1000 turns of No. 32 copper magnet wire, wound from a total length of wire measuring 200 feet. What current will a 6-volt storage battery establish in the coils? What is the total magnetomotive force created by this current?

Prob. 16-10. If the 6-volt battery of Prob. 15-10 were replaced by an 8-volt battery, what total magnetomotive force would the magnets create? How many times as strong is the magnetomotive force with the 8-volt battery compared with the strength of mmf created with the 6-volt battery?

Prob. 17-10. Five single-stroke gongs are to be placed on five different floors of a commercial building to summon a janitor to the superintendent's head office. Show the diagram of connections if there is to be a single push button in the head office.

Prob. 18-10. Each of three floors in a factory has an alarm push button with one normally open contact. When any one of the buttons is depressed momentarily, a vibrating-stroke bell in the guard office is to begin ringing and is to continue ringing until a reset button is depressed in the guard office. The circuit includes a relay having two normally open contacts. Show a complete diagram of connections for operation from a single 12-volt d-c source.

Prob. 19-10. Show a diagram of a circuit with a normally open float switch for ringing a bell when water in a tank reaches a specified level.

Prob. 20-10. When the door of a small retail store is opened, a buzzer is to sound at the back of the shop. Show a diagram of connections, and describe the switching device needed at the door.

Prob. 21-10. Through how many feet of No. 18 bell wire can a buzzer requiring 0.5 ampere at 4 volts be rung from a 6-volt d-c source?

Prob. 22-10. An electric gong with an operating current of 5 amperes is to be controlled from a push button 1000 feet away. Draw a diagram showing the circuit if the power is supplied from a power source located near the push button.

Prob. 23-10. The operating voltage for the gong of Prob. 22-10 is 15 volts when its current is 5 amperes.
(a) What power is consumed by the bell itself?
(b) What power is lost in the line wires if these are No. 18 copper wires?
(c) What power is lost in the line wires if these are No. 14 copper wires?
(d) What is the source voltage for parts (b) and (c)?
(e) Would there be any saving in power if the d-c source were placed near the gong rather than near the push button?

Prob. 24-10. From the results of Prob. 23-10, the power lost in the line wires is much greater than the power needed for the gong itself. Draw a diagram of connections using a sensitive relay to control the current in the gong. The local circuit through the gong is to contain 25 feet of No. 14 copper wire. What source voltage does this connection require?

Prob. 25-10. What voltage is applied to the relay of Prob. 24-10? If the resistance of the coil of the relay is 50 ohms, what is the current in the push button circuit when No. 18 wire is used?

Prob. 26-10. What is the efficiency of transmission from the source to the gong in part (c) of Prob. 23-10? In Prob. 24-10?

Prob. 27-10. Show on a diagram how an electric lock opener and a doorbell can be operated from a single d-c source.

Prob. 28-10. Design a three-position switch to control the lamps of an automobile as follows: Position 1, all lights off; 2, parking and tail lamps on; 3, head and tail lamps on. Sketch the complete circuit diagram.

11
D-C Machines

1. Reversibility. Any electric motor may be run as a generator, or vice versa. If electric power is generated outside the machine and brought to it, the machine is said to be a *motor* if this power sets the machine in motion and causes operation of other machinery attached to the machine. On the other hand, if mechanical power from some outside source is applied to the electric machine, the machine is called a *generator* if it can then make electric power available at its terminals for use in operating other electric circuits.

2. Basic Principles. Motor Action and Faraday's Law. Because any electric machine can be either a motor or a generator, every electric machine has the same two basic principles at work inside. The first of these principles, *motor action,* was introduced in Chapter 4. Whenever a current-carrying wire is immersed in a magnetic field, there is a force that tries to move the wire. This force is known as the force of motor action. It is the same force that causes the bobbin of an indicating instrument or meter to turn through the magnetic field of the meter when there is a current in the wires wound on the bobbin.

The second basic principle of electric machines is a relationship between voltage and speed, known as *Faraday's law,* in honor of the discoverer. Figure 192 shows an arrangement of an experiment that will demonstrate Faraday's law. The bar magnet, with its north pole marked N, is held stationary. The wire marked *x-y* is moved quickly from left to right across the end of the bar magnet. As long as the wire is in

224 Essentials of Electricity

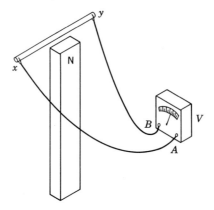

Fig. 192. When the wire is moved across the face of the magnet's pole N, the voltmeter indicates that a voltage is established in the wire.

motion through the magnetic flux lines, the electric charges in wire x-y are moving through the magnetic field. These charges in mechanical motion may be thought of in a somewhat similar fashion as the charges in motion that constitute the current involved in motor action. Just as a current-carrying wire experiences a force on its side in a magnetic field, the charges inherent in wire x-y of Fig. 192 are pushed on one side by the combined action of the field and the motion of the wire. Considering the internal charges of the wire as positive, in accordance with the normal definitions of circuit theory, we find that the positive charges are crowded toward the x end of the wire, the y end becoming negative.

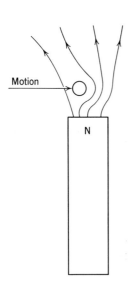

Fig. 193. The direction in which the flux lines might wrap around a wire indicates the polarity of the Faraday voltage.

Another view of the magnet and the wire is shown in Fig. 193. The x end of the wire is nearer the reader than the y end. Recall that the sense of magnetic flux lines has been defined as being directed *out* of a north pole; then the *right-hand rule* may be employed to determine the polarity of x (positive as stated above). Place the fingers of the right hand in the direction of flux lines leaving the north pole of the bar magnet. As wire x-y moves across the top of the magnet, consider that the flux lines (and your fingers) wrap around the wire. The thumb of the right hand will then point to the end of the wire that is positive. If terminal A is the positive terminal of the voltmeter in Fig. 192, the Faraday voltage created by the motion of the wire will establish a current in the circuit from x to A and B to y and back to x, and the voltmeter will read upscale.

If the motion of the wire in Fig. 193 were from right to left, the voltmeter would read downscale. If there is no motion between the wire and the magnetic flux, the generating action stops and there is no Faraday voltage. Further experimenting shows that twice as much voltage is developed if the same wire is moved twice as quickly through the flux lines. If an electromagnet sets up the flux lines, the strength of the magnetic field might be doubled by a sufficient increase in the current in the electromagnet. Faraday's law also includes the effect of increasing the magnetic field strength: for a given wire and a fixed speed, doubling the field strength also doubles the Faraday voltage. Large voltages may be generated by moving many wires in series very rapidly across the faces of very strong electromagnets.

In many practical electric machines, the speed of the wires moving through the magnetic lines is constant. However, the strength of the magnetic lines usually varies from one part of the electric machine to another. Therefore, the Faraday voltage is usually an alternating voltage, changing with time. The study of such voltages involves a study of time rates, which have not been important in this book. Although the Faraday voltages in d-c machines possess this alternating nature, the method of connecting the moving wires to the external circuit produces a voltage that is a steady direct voltage at the terminals.

In summary, then, Faraday's law states the relationship between voltage and speed. The magnitude of the voltage produced in a circuit moving through flux lines is proportional to the product of three factors: (1) the speed of the moving wires; (2) their total length (number of wires times unit length); and (3) the strength of the magnetic flux. The sense of the Faraday voltage is found with the aid of the right-hand rule.

3. Basic Parts of Electric Machines. Most electric machines may be subdivided into two basic sections: (1) the *stator,* or stationary part of the machine, and (2) the *rotor,* or movable portion. In most designs, the stator is the outer member of the machine, and the rotor is inside. For ordinary d-c machines, the stator comprises several electromagnets arranged to create flux lines that pass through the iron yoke of the stator frame, into the iron of the rotor across the relatively small air gap between the stator and rotor. The arrangement of a typical d-c machine can be inspected in Fig. 194.

226 Essentials of Electricity

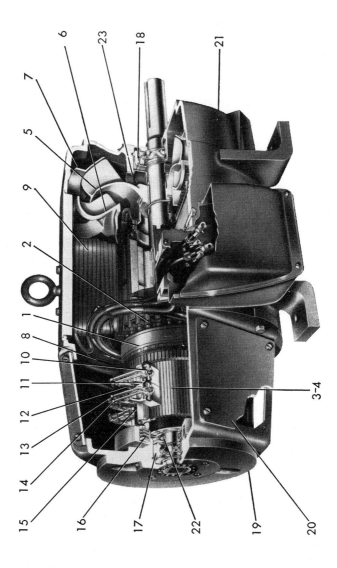

11: D-C Machines   227

reference number	description of part
1	armature complete with shaft and blower
2	armature coils
3	commutator
4	assembled segments
5	fan
6	air diffuser
7	air shield
8	shunt or compound coil
9	commutating coil
10	brushes
11	brushholder
12	spring
13	clip—tension adjusting, short finger
14	clip—tension adjusting, long finger
15	spring retaining clip
16	rocker ring
17	ball bearing—front
18	ball bearing—rear
19	bracket—front
20	bracket covers
21	bracket—rear
22	bearing cap—inner, front
23	bearing cap—inner, rear

Fig. 194. Cutaway view of a d-c machine. *Westinghouse Electric Corporation.*

228 Essentials of Electricity

Fig. 195. Rotor structure (armature) of the machine shown in Fig. 194. *Westinghouse Electric Corporation.*

Because the stator coils establish the principal magnetic field for the machine, the collection of these coils is usually referred to simply as the *field* of the machine, or the *field winding*. In a d-c machine, then, the stator winding is called the field winding.

Embedded in the iron of the rotor are other wires, in which currents will react with the magnetic flux lines to produce the force of motor action. These wires are insulated from each other and from the iron slots in which they lie. In Fig. 195, groups of wires can be seen taped together, lying in the slots of the rotor iron. Because such a construction of wires is thought of as being "armored" with iron, the rotor structure of a d-c machine is usually called the *armature* of the machine. Therefore, in most d-c machines the armature winding is the rotor winding.

At the left end of the armature shown in Fig. 195 is a device known as the *commutator*. Many slender bars of copper are mounted around the shaft of the machine, insulated from each other and from the shaft and the iron of the rotor. There is frequently the same number of copper bars as the number of slots in the rotor iron. This assemblage of copper bars makes up the commutator. Figure 194 should also be inspected to determine the location of the commutator.

Bearing on the commutator is a set of graphite blocks, called *brushes*. Heavy conductors joined to the brushes provide the connections for the armature terminals of the machine. The wires within the armature winding itself are joined to the commutator bars in a kind of woven fashion. The combination of the method of winding, the commutator, and its brushes provides sliding connections to the rotor winding, permitting the armature to rotate continuously. Most important, however, is the fact that the commutator is the device that causes the time-changing Faraday voltages of each of the rotor wires to appear as a constant direct voltage at the armature terminals. This behavior is known as *commutator action,* or *commutation.* Because the commutator also switches heavy currents to the various wires in the armature winding while the rotor is in motion, successful commutation means the performance of this switching of heavy currents without excessive or dangerous sparking. Most d-c machines have small electromagnets located between the main field electromagnets. Because the main field windings create the principal magnetic *poles* in the machine, the large electromagnets are called the *main poles.* The smaller electromagnets, known as *interpoles,* create auxiliary zones of magnetomotive force in the vicinity of the wires in which the commutator is switching the heavy current. Without interpoles, commutation is usually not very satisfactory. Therefore, although interpoles do not directly influence the basic motor action and Faraday voltage of d-c machines, interpoles are needed to make commutation practical.

4. Magnetic Paths. A d-c machine could be made with a permanently magnetized stator structure as shown in Fig. 196, where the rotor has been removed. The stator may be considered to be a horseshoe magnet like the one shown in Fig. 49 of Chapter 4. The complete, closed loops of flux are shown both inside and outside the iron in Fig. 196. Note that the lines come out of the north pole and enter the south pole, go through the magnetized iron yoke back again to the north pole. When the armature is inserted in its proper place between the two stator poles, the strength of the magnetic field will be much greater because the permeability of the iron in the armature is much greater than that of air.

Because most machines are cylindrical in cross section, the closed loops of the magnetic flux usually form several parallel paths within the machine. Figure 197 shows the magnetic paths in a common form of stator. Note that the flux lines come out of a north pole and go into a south pole, and note that north poles and south poles alternate around the frame.

When the electromagnets project out of the stator yoke as in Fig.

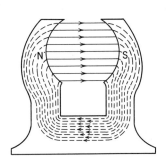

Fig. 196. Flux paths in a permanently magnetized 2-pole stator (field) structure.

197, the machine is said to have *salient poles*. Because most d-c machines have this salient stator construction, it is usually possible to tell how many magnetic poles a machine has by counting the number of main salient projections. However, the machine of Fig. 197 is a 4-pole machine because there are four main bunches of magnetic flux in the field pattern, the count of four being obtained in the air gap between the rotor and stator. A similar tallying for the permanent-magnet machine of Fig. 196 shows that that machine is a 2-pole machine.

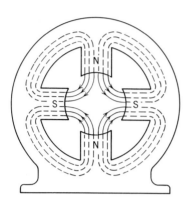

Fig. 197. Magnetic lines in a 4-pole machine.

Prob. 1-11. Draw a 2-pole motor frame with the cylindrical shape of Fig. 197, showing the complete paths of the magnetic flux lines.

Prob. 2-11. Draw a 6-pole motor frame showing the complete paths of the magnetic lines.

Prob. 3-11. Draw an 8-pole generator frame showing the complete paths of magnetic force lines.

Prob. 4-11. Sketch the four salient projections of the frame shown in Fig. 197. Arrange flux lines so that two adjacent pole pieces are north poles, and the other adjacent pair are south poles. (This arrangement is found in generators for certain systems of automation.) How many poles does the generator have?

5. Types of Field Coils. Except for small permanent-magnet machines, the iron used in d-c machines is usually soft iron. Coils of wire around the salient polar projections of the stator establish the principal magnetic flux of the machine. Because most d-c machines obtain current for these field coils from the line voltage at which the machine is to operate, these field coils contain many turns of relatively small wire. There is usually one such main field coil per salient projection, and these main windings are joined in series with each other. Because this connection of coils will be operated as a series string of resistances from the line voltage, the string of main coils is actually in parallel (in shunt) across the line wires. Therefore, the main field winding of a d-c machine is usually called the *shunt field*.

Each interpolar projection of a machine also is wound with a coil, and these coils are joined in series with each other. Since these coils provide a corrective mmf to improve commutator action, the current to establish this correction must be the large current of the rotor wind-

11: D-C Machines 231

ing. Therefore, the wire in the interpole winding must be large, for the interpole winding will be connected in series with the armature leads to the brushes of the machine. Both the main field and the interpole winding of a machine can be seen in Fig. 198.

Certain desirable characteristics of d-c machines can be further enhanced by the addition of corrective windings placed around (and usually on top of) the main shunt-field windings. If there is one such added coil around each polar projection, the coils are again joined in series, this time to be added essentially in series with the armature current. Therefore, the wire in these additional coils must be large to carry the rotor current safely, and this corrective winding is known as the *series field* of the machine. Figures 194 and 198 show the coils of the series field arranged on top of the shunt-field coils. All field coils are usually prewound, slipped in place on the correct pole pieces, and then joined together electrically.

Fig. 198. Field coils in place on the stator structure of the machine shown in Fig. 194. *Westinghouse Electric Corporation.*

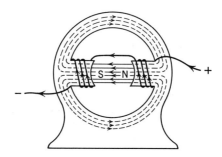

Fig. 199. Winding sense for field coils of a 2-pole machine.

A schematic illustration of the shunt-field winding for a 2-pole machine is shown in Fig. 199. The current and flux directions should be checked with the right-hand rule. When a machine must be rewound and rewired, a compass and the right-hand rule are extremely helpful in assuring that the correct magnetic flux pattern is established with the new winding.

Prob. 5-11. Draw a sketch of a 6-pole motor. Add the lines of magnetic flux through the pole pieces and the stator yoke. Show field coils around the pole pieces. Connect the coils in series properly so that a single current in all of them creates the correct sense of poles in the salient stator pieces, in accordance with the right-hand rule.

Prob. 6-11. Put field coils on the generator of Prob. 3-11. Connect these in series properly in accordance with the right-hand rule so that a single field current will establish the correct pole polarities.

6. Types of Connections. When the principal magnetic flux of a d-c machine is established from a current from one electric source, and when the current in the armature is established from a second source, the machine is said to be *separately excited.* Although some motors in automatic-control schemes are separately excited, the term more often applies to generators. When a d-c generator is connected so that the line voltage it produces from its armature also supplies the current for its main magnetic field, the generator is said to be *self-excited.* Because most motors receive both their field and armature currents from the same line wires, motor connections are classified the same way as self-excited generator connections.

Three classes of self-excited connections should be distinguished:

Shunt. The shunt field of the machine, connected in parallel with the rotor-circuit terminals, provides the magnetic flux for the machine. The stator-winding current is then usually only a small fraction of the line current.

Series. A special main-field winding is constructed of large wire so that the armature current can create all of the flux of the machine. In a series machine, the line current equals the field current, which also equals the armature current.

Compound. The magnetic flux of the machine is provided from both a shunted winding and a series winding. Most compound machines are essentially shunt machines with a corrective component of magnetomotive force provided by the series field.

Two subdivisions should be noted for compound machines. When the corrective component of mmf *adds* to the main shunt-field mmf, the machine is said to be *cumulatively compounded*. A machine with the windings connected so that the correction of the series field is *subtractive* is said to be *differentially compounded*. The differentially compounded motor finds almost no use, and the differentially compounded generator has only special applications. Therefore, most compound machines are connected for cumulative operation; care must be taken in making connections to see that the cumulative condition is achieved.

7. Number of Brushes. When large currents must be switched by the commutator, a single brush placed at the appropriate location on the copper bars may not be able to commutate the current. For such machines, the commutator bars are longer in axial length along the shaft to provide more area for switching the large current. However, a single long brush may provide only a few points of contact because of mechanical vibration, making commutation poor or dangerous. Therefore, long commutators for high currents have a *set* of brushes at each brush location, thereby insuring smoother riding of the contacting surfaces.

The ordinary d-c machine has the same number of sets of brushes as it has poles, not counting the interpoles. Note that the generator in Fig. 200 has six poles and six sets of brushes. Alternate sets of these brushes are positive, and may be joined together. With a similar joining of the negative sets, the two available terminals from the machine provide a connection of essentially three generating circuits in parallel. Multipole machines are usually high-current machines. Figure 201 shows schematically how the brush sets of Fig. 200 are connected to provide two terminals for the generator.

8. Commutating Poles. The interpoles, or *commutating poles*, do not directly assist in the power conversion in the armature. However, they do serve to keep the brushes from sparking on heavy loads or high speeds. The coils on the interpolar projections are always connected in series with the armature, as noted above, so that their mmf depends

234 Essentials of Electricity

Fig. 200. Brush rigging on a 6-pole generator. *General Electric Company.*

upon the armature current. With appropriate design of interpoles, machines can be made to reverse quickly from full speed in one direction to full speed in the opposite direction, without sparking at the brushes. Figure 202 shows a sketch of a 2-pole machine in which one of the interpole windings has been assembled.

The polarity of the interpoles can be found as follows: Determine the polarity of the main poles and the direction of rotation of the armature.

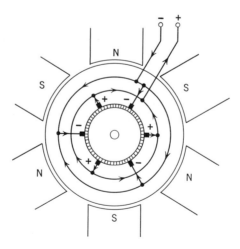

Fig. 201. Brush connections for the 6-pole machine of Fig. 200.

11: D-C Machines   235

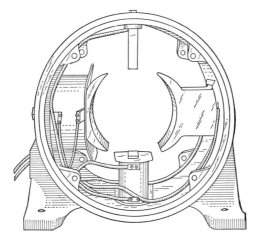

Fig. 202. Frame of a 2-pole machine with stator windings partly assembled.

Then, place your hand on one pole after another to make a complete traverse around the circumference of the machine in the same direction as the direction of rotation. If the machine is a generator, any interpole has the same polarity as the next main pole *following* after the interpole in your traverse. If the machine is motoring, the polarity of any interpole is the same as that of the main pole passed *just before* the interpole in your traverse.

Note in Fig. 203 that, if we start with the main north pole at the top of the diagram and go around the frame clockwise (the direction in which the armature is rotating), the first interpole is a south pole. This interpole has the same polarity as the next main pole that follows after it in the clockwise traverse of the machine.

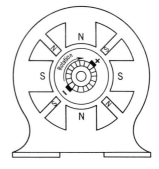

Fig. 203. Commutating poles on a 4-pole generator.

Prob. 7-11. Draw coils on the main and commutating poles of the generator of Fig. 203. Connect the interpole windings properly in series with the positive armature brush (the second pair of brush sets has been omitted for simplicity). Connect the shunt-field coils properly to a source for separate excitation.

Prob. 8-11. In Fig. 203, keep the polarities of the main poles the same as shown. Let the direction of rotation be counterclockwise. Then the armature voltage at the brushes will reverse polarity, because the speed in Faraday's

236 Essentials of Electricity

law is now in the opposite sense. Determine the polarity of the interpoles for this new operation.

Prob. 9-11. Repeat Prob. 7-11 for the conditions of Prob. 8-11.

Prob. 10-11. Show that if the interpole connections have been made correctly for Prob. 7-11, the same connections are also correct for Prob. 9-11.

9. Equivalent Circuits for Basic Parts. It should be clear from the preceding description of the basic parts and principles of operation that understanding even simple machines involves a fair number of relationships. However, there is a possibility of combining these relations into equivalent-circuit representations, in the manner of the equivalent circuits for battery-cell combinations in Chapter 8. Calculations may then be made on the equivalent circuit as an estimate of what the actual behavior of the machine will be like.

For each of the stator windings, representation with a single resistance is adequate for direct current. Since the shunt-field winding usually has many turns of relatively small wire, its representation is often a resistance of 10's or 100's of ohms. A series-field winding with a few turns of large wire usually has only a fraction of 1 ohm for its resistance. Interpole windings often have even smaller resistance than do series windings. Figure 204 shows a standard method by which stator-winding terminals are designated, and the equivalent-circuit representations as resistances.

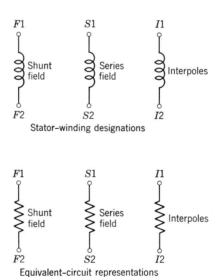

Fig. 204. Terminal markings and d-c equivalent circuits for stator windings.

In the armature there is a more complicated action. The presence of the switching action of the commutator makes it possible to think of each wire in the rotor winding as a battery cell. Each wire then has its own internal Faraday voltage due to its speed as it passes the main poles of the machine. Although large conductors are used in the rotor winding, each rotor wire has a small resistance that could be compared to the small internal resistance of a battery cell. Therefore, when measurements are made at the two terminals of the armature connections, the rotor of any d-c machine may be represented by an equivalent internal voltage in series with an equivalent internal resistance. This combination is essentially a Thevenin representation of the armature circuits. The commutator and the brush-set connections provide many series-parallel arrangements of the internal wires in the rotor winding. In any machine, then, these series-parallel arrangements may always be reduced with the aid of Thevenin's theorem to a single series circuit. Figure 205 shows a standard marking for the armature terminals and the Thevenin-equivalent representation. Usually the internal voltage is nearly equal to the terminal voltage of the machine, and the internal resistance is a fraction of 1 ohm.

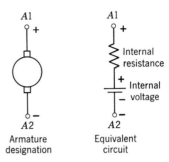

Fig. 205. Terminal markings and d-c equivalent representation for the armature circuit.

10. Separately Excited Generator. With the symbols from the preceding section, Fig. 206 shows the connections for a separately excited generator. The equivalent-circuit representation is shown in Fig. 207. Note that the field current exists in one circuit, and that the armature current is in a completely separate circuit. Such an arrangement is of the utmost importance when the polarity of the load voltage must be controlled.

11. Shunt-Generator Connections. The essential feature of a shunt generator is that its main field is in parallel with its armature circuit. In order to permit an easy adjustment of the shunt-field current (and ultimately the terminal voltage of the generator), an adjustable resistor called the *field rheostat* is inserted in series in the shunt branch. This addition does not alter the basic parallel connection, however. Figure 208 shows the symbolic connections, and Fig. 209 illustrates the equivalent-circuit representation. Note that the interpoles have been

238 Essentials of Electricity

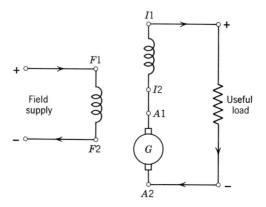

Fig. 206. Connections for a separately excited generator.

included within the armature terminals. Most machines do not have the interpoles brought out separately, except for testing. Therefore, the resistance shown in Fig. 209 for the armature circuit includes the internal resistance of the rotor winding, the resistance of the interpole winding, and an additional resistance to represent the sliding contact of the brushes on the commutator.

Usually the leads of a generator are banded with number designations, or they terminate on numbered posts in a connection box. Figure 210 indicates how the terminals of a shunt generator might appear on a panel. Figure 211 shows what the wiring layout would be for the connection of Figure 208. Note that points x and y in Fig. 211 are included between points $F1$ and $F2$ in Fig. 208. When the rheostat is mounted with the machine it controls, and when the terminals of the

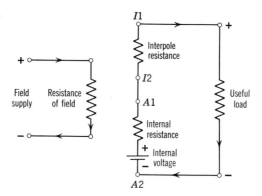

Fig. 207. Equivalent circuit for a separately excited generator with load.

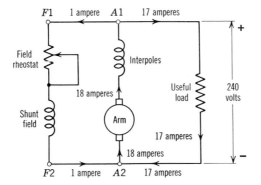

Fig. 208. Connections for a shunt generator.

rheostat are brought out at the machine terminals, the labeling of Fig. 208 is often found. Otherwise, terminals $F1$ and $F2$ refer only to the field of the machine.

Prob. 11-11. The terminals of a shunt machine were found to be labeled on a panel as shown in Fig. 212. (The armature circuit is between A and $A1$; the field terminals are F and $F1$.) Show connections to a field rheostat and to a load, in the manner of Fig. 211.

12. Shunt-Generator Build-Up. Even when there is no current in the field coils of a machine, there is always a small amount of residual magnetism from the last time the machine was used. When the prime mover that supplies mechanical power to a shunt generator gets the shaft turning, the armature conductors pass through a small number of residual flux lines. This action produces a very low internal voltage, in accordance with Faraday's law. If the load is not connected to the generator, this very low internal voltage can establish a very small field

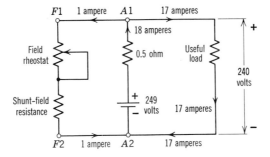

Fig. 209. Equivalent circuit for a shunt generator with load.

240 Essentials of Electricity

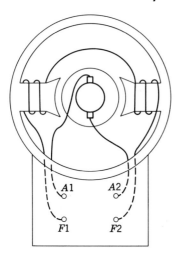

Fig. 210. Possible panel connections for a d-c machine.

current, as the circuit of Fig. 209 indicates. This small field current will, in turn, produce an increase in mmf and in the flux of the main poles. Since this flux is the active flux in Faraday's law, the internal voltage of the armature circuit increases slightly. Each increase in internal voltage keeps increasing the field current, causing an action known as *building up*. If a voltmeter has been connected to the armature terminals before the prime mover is started, the pointer will be observed to increase from a very low value to a value near the full voltage of the generator. This building-up process usually occurs in from 10 to 30 seconds after the generator is started.

The question may be asked: Why does the building-up process finally stop? It would appear that doubling the internal voltage would double the field current, etc. However, when the generator is producing an internal voltage near its full-voltage rating, the relationship between field current and magnetic flux is not a proportional one, as are the relationships in the equivalent circuit. It is this lack of proportionality in the

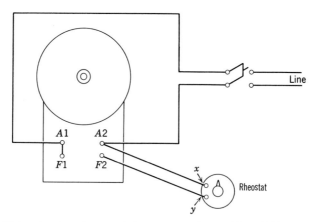

Fig. 211. External connections needed to operate the machine of Fig. 210 as a shunt generator.

magnetic behavior of the machine that determines a definite full-voltage operating condition. However, after the normal-voltage range is reached, considerable adjustment in the terminal voltage of the shunt generator may be made by changing the field current with the field rheostat. Increasing the field-circuit resistance lowers the field current, weakening the magnetic field, and lowering the internal voltage generated. Since Faraday's law also includes speed as a factor, a higher internal voltage can be generated if the speed of the prime mover is increased.

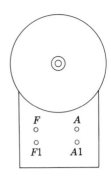

Fig. 212. These terminal markings are not standard.

Sometimes a self-excited shunt generator does not build up when the prime mover is started. This difficulty happens if the sense of the internal voltage is such that the field current it creates sets up an mmf in opposition to the original residual magnetism of the machine. This condition can sometimes be noted with a voltmeter across the unloaded generator terminals. If the generator is operated with all of the resistance of the rheostat added into the field circuit, a very low voltage will be observed. Then if the resistance is gradually cut out of the field circuit, the voltmeter may be observed to drop 1 or 2 volts further. This drop indicates that the increased field current is further reducing the internal generated voltage, because the residual flux and the flux due to the field mmf are in opposition.

The remedy for this trouble is to reverse the field connections. Refer to Fig. 211 for the original connections. If the machine does not build up with this arrangement, connect $F1$ to $A2$, and join $F2$ (through the field rheostat) to $A1$. Very rarely a machine may not build up because it has no residual magnetism. The shunt field may then have to be disconnected and excited from any available source. For completely isolated generators, storage batteries or dry cells may have to be employed to leave the machine with sufficient residual magnetism for the building-up process.

13. Compound-Generator Connections. When a load is connected to the terminals of a shunt generator, the increase in armature current needed to supply the load causes a drop in terminal voltage below that of the internal generated voltage. This action is not different from the internal drop that occurs in a battery cell under load. However, the lowered terminal voltage of the shunt generator also reduces the field

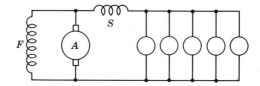

Fig. 213. Diagram of generator connections for short-shunt compounding.

current, causing a chain of events that results in a drastically lowered terminal voltage when a new magnetic equilibrium is reached.

A cumulatively-compounded generator does not have this fault of a shunt generator. The load current (or sometimes the armature current) is passed in series through a corrective series winding, causing an increase in mmf on the main poles at the same time that the increased armature current tends to lower the terminal voltage of the generator. By a suitable choice of the number of turns in the series winding, the internal voltage of the machine can be increased at its full-load point sufficiently just to make up for the full-load drop in the internal resistance of the equivalent circuit of the armature. In such a machine, the no-load and full-load terminal voltages are the same, and the machine is said to be *flat-compounded*.

Figure 213 shows a symbolic connection diagram for a compound generator, coils S being the series winding, and coils F the shunt winding. The circuit of Fig. 213 shows the load current in the series winding, an arrangement that shunts the main field only across the armature terminals. Such a connection is called *short-shunt compounding*. If it is desirable to have only the armature current in the series winding, the connection of Fig. 214 may be arranged. Since the main-field terminals span a larger circuit in this form of compounding, the connection is described as *long-shunt compounding*. Flat-compounded behavior may be obtained with either connection.

The mmf created by the series winding is normally not more than 20

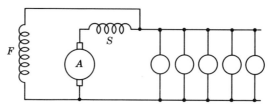

Fig. 214. Diagram of generator connections for long-shunt compounding.

to 30% of the mmf of the shunt winding. When large proportions are chosen for the series-winding ampere-turns, the action of the generator becomes more like that of a series machine. Series generators have special characteristics that are occasionally useful in automatic-control systems.

14. Torque and Power. When a force of 10 pounds pushes a load 2 feet across a level floor, we say that 20 *foot-pounds* of *work* has been accomplished. The force acts in a straight line, and the work is accomplished along that straight line. In rotating machines, however, a point on the shaft moves in a circle around the center of the shaft. For this rotary motion, the force involved in the turning effort depends on how far from the center of the shaft the force is measured. For example, an automobile engine might have a brake drum attached that has a radius of 1 foot. Measured at the circumference of this brake drum at a certain speed, the force developed by the engine is 220 pounds. However, if this drum were removed and a new drum were attached with a radius of only $\frac{1}{2}$ foot (or 6 inches), the same engine would develop a force of 440 pounds acting at a distance of 6 inches from the shaft.

Therefore, for rotary motion we need a new term for describing the turning effort. This term is *torque,* and it is measured in the number of pounds of force that is exerted times the radial distance at which the force is exerted. For the automobile engine described here, we would say that it develops a torque of 220 pound-feet, this value being the same whether we multiply 220 pounds by 1 foot, or 440 pounds by $\frac{1}{2}$ foot. Notice that the British unit of torque is *pound-feet,* stated in the order of pounds first and feet second to avoid confusion with the unit of work, i.e., foot-pounds. (In the metric system, as noted in Chapter 7, the unit of work is the joule. The unit of torque is the newton-meter, meaning the force in newtons measured one meter away from the center of the rotating shaft.)

As noted in Chapter 9, the circumference of a circle is π times its diameter. Since the diameter of a circle is twice the radius, we may find the circumference of a circle by multiplying the radius by 2π. Thus, for the automobile engine described above, the force of 220 pounds on the drum of 1-foot radius acts through a distance equal to the circumference of the drum every time the drum is turned one revolution. The work per revolution is thus seen to be 2π times the torque of the engine, or: $2 \times 3.14 \times 220 = 1382$ foot-pounds per revolution, for the conditions described. Since the torque always may be expressed as the force at a distance of 1 foot from the center of a rotating shaft, the work per revolution of the shaft may always be expressed as 2π times the torque.

244 Essentials of Electricity

Recall from Chapter 7 that 1 horsepower is defined as the rate of doing work equal to 33,000 foot-pounds per minute. If we know that the automobile engine turns at 3000 revolutions per minute (usually abbreviated rpm) when it is developing its torque of 220 pound-feet, we know that its work per minute is: (work per revolution) times (revolutions per minute). For this example, the rate of doing work is: (1382 foot-pounds per revolution) times (3000 rpm) equals: 4,146,000 foot-pounds per minute. If the rate of doing work is now divided by 33,000 foot-pounds per minute, the energy rate of the engine will be stated in horsepower:

$$\frac{4,146,000}{33,000} = 126 \text{ horsepower}$$

The energy rate or power of a rotating machine, whether mechanical or electrical, may be expressed as indicated by the preceding calculations. This process may be summarized as:

$$\text{horsepower} = \frac{2\pi \times \text{torque in pound-feet} \times \text{speed in rpm}}{33,000}$$

Since the essential part of the production of useful power involves the torque multiplied by the speed of the shaft, the other numbers are only needed to get the answer into the form of horsepower. Numerical labor can be saved if the numerator of this ratio is divided by 2π, and the denominator is also divided by 2π to keep the balance. Then the calculation for horsepower becomes:

$$\text{horsepower} = \frac{\text{torque in pound-feet} \times \text{speed in rpm}}{5250}$$

The number 5250 equals $33,000/2\pi$. For the example of the automobile engine given above:

horsepower =

$$\frac{220 \text{ pound-feet} \times 3000 \text{ rpm}}{5250} = \frac{660,000}{5250} = 126 \text{ horsepower}$$

We should note that the calculation of horsepower from torque and speed is essentially a product, just as the calculation of watts is the product of volts and amperes. In fact, we could write the product thus:

$$5250 \times \text{horsepower} = \text{torque in pound-feet} \times \text{speed in rpm}$$

In Chapter 7, we saw that there were three forms for the power equation for electric units, just as the product of Ohm's law has three forms. In

11: D-C Machines   245

a similar fashion, the torque-speed product above can be written in two other forms. The first is useful for finding the torque when the horsepower and speed are known:

$$\text{torque in pound-feet} = \frac{5250 \times \text{horsepower}}{\text{speed in rpm}}$$

The second is helpful in speed calculations:

$$\text{speed in rpm} = \frac{5250 \times \text{horsepower}}{\text{torque in pound-feet}}$$

Prob. 12-11. A certain automobile engine has an advertised torque of 380 pound-feet at 2400 rpm. What horsepower does it develop under these conditions?

Prob. 13-11. The engine of Prob. 12-11 is advertised as capable of producing 265 horsepower at a speed of 4400 rpm. What torque does it produce at this advertised condition?

Prob. 14-11. An electric motor is rated to produce 15 horsepower at 1840 rpm. What torque does it produce?

Prob. 15-11. The second hand on a timer requires a torque of 0.5 ounce-inches (a force of 0.5 ounce acting at a distance of 1 inch from the center of the shaft) to operate accurately. If the dial reads from 0 to 60 seconds for one revolution of the second hand, what horsepower must the timer motor supply to drive the hand?

15. Energy Conversion. In either a motor or a generator of the d-c kind, mechanical energy is converted to electrical energy (or vice versa) in the armature. Figure 215 repeats the equivalent circuit of Fig. 207 for a separately excited generator, with some numbers added to make the

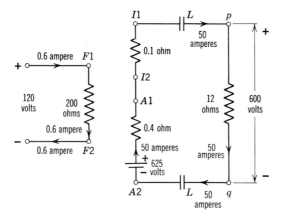

Fig. 215. Shunt generator supplying power through a load contactor with contacts marked L.

description of the machine more specific. Two normally open contacts, marked L, are shown in series with the load. The load current of 50 amperes sets up a voltage of 600 volts across terminals p and q of the load when the load contactor closes the contacts marked L. Energy conversion occurs inside the armature, where the internal voltage of 625 volts sets up the load-circuit current of 50 amperes. The rate at which energy is being converted from mechanical to electrical form may be expressed in terms of the armature internal power:

$$\begin{aligned}\text{internal power} &= \text{internal voltage} \times \text{internal current} \\ &= 625 \text{ volts} \times 50 \text{ amperes} \\ &= 31{,}250 \text{ watts} = 31.25 \text{ kilowatts}\end{aligned}$$

This calculation shows that inside the armature energy is being converted from mechanical form to electrical form at the rate of 31.25 kilowatts. In Chapter 7, we saw that power could be expressed either in the metric units of kilowatts, or in the British units of horsepower. As shown there, 1 kilowatt = 1.34 horsepower; i.e., there are more units of horsepower in describing a given rate than there are units of kilowatts, since the kilowatt is the larger unit. The internal power of the generator of Fig. 215 is:

$$\text{horsepower} = 1.34 \times 31.25 \text{ kilowatts} = 41.9 \text{ horsepower}$$

This calculation of internal horsepower for a generator is more than an exercise in converting units: this value in horsepower represents mechanical power supplied from the prime mover as it turns the generator. The electrical power in kilowatts expresses the rate at which power is being supplied to the armature circuit by the energy-conversion process going on in the armature.

For further insight into this process, let us assume that the speed of the generator is 2800 rpm. Further, let us ignore any mechanical losses in the bearings holding the shaft of the generator, and in the rotation of the relatively rough armature through the surrounding air. (Although these losses are small, the prime mover must supply them mechanically; they are usually called friction-and-windage losses.)

With the calculation for torque from the preceding section, we find:

$$\text{torque} = \frac{5250 \times \text{horsepower}}{\text{speed in rpm}} = \frac{5250 \times 41.9}{2800} = 78.5 \text{ pound-feet}$$

Therefore, with the separately excited generator of Fig. 215 turning at 2800 rpm and supplying 50 amperes to its load, the prime mover must develop 78.5 pound-feet of torque on the shaft of the generator. From another point of view the 50-ampere current in the armature conductors develops forces of motor action with the flux lines of the main magnetic field. In turn, these motor forces act to oppose the effort of the prime mover to keep the generator turning. The opposition of the armature to turning is exactly what requires the torque from the prime mover, in order that electric power can be made available in the armature circuit. The force relations are sketched in Fig. 216.

In review, then, this example shows the action of the two basic principles of electric machines, both occurring simultaneously in this separately excited generator. When the armature is turned by the prime mover at 2800 rpm, and when the main field winding sets up the necessary magnetic flux, the Faraday voltages in the separate wires of the rotor winding produce the equivalent internal Thevenin voltage of 625 volts. When there is a load current of 50 amperes in the armature circuit, the prime mover must increase its torque by 78.5 pound-feet to supply the additional mechanical power that is being converted into the internal electrical power of 31.25 kilowatts.

Prob. 16-11. The generator of Fig. 215 is to be reconnected as a shunt generator. Therefore, a resistance must be added in series with the main field to keep the field current the same as before. If the load current and voltage remain the same as in Fig. 215, and if the internal voltage remains essentially at 625 volts, what must the size (in ohms) of the added resistance be? Draw the new circuit diagram; label clearly the size and direction of the current in each part of your diagram.

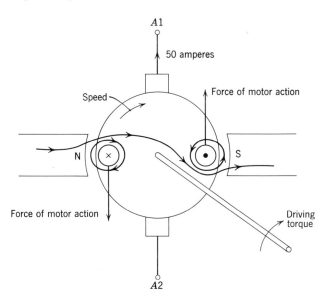

Fig. 216. Torque and force directions in a generator.

Prob. 17-11. An accident causes the load contactor to open contacts L shown in Fig. 215. A governor holds the speed of the prime mover at 2800 rpm. Assume that the internal voltage remains constant at 625 volts. What is the voltage between load terminals p and q? What is the voltage between the machine terminals $I1$ and $A2$? What torque is supplied by the prime mover?

Prob. 18-11. Repeat Prob. 17-11 for the circuit of Prob. 16-11. Draw the new circuit diagram; label clearly the size and direction of the current in each part of your diagram.

Prob. 19-11. The speed of the separately excited generator of Fig. 215 is reduced to 1400 rpm. Because of the nonproportional nature of its magnetic circuit, the internal Thevenin voltage becomes 350 volts, with the field current still at 0.6 ampere. If the 12-ohm load remains connected to the armature, what is the new load current? What is the drop in the inside of the machine (armature and interpoles)?

Prob. 20-11. What voltage appears between terminals $I1$ and $A2$ of the machine described in Prob. 19-11 when the load contactor disconnects the 12-ohm load by opening its contacts L?

16. Shunt Motor.

If the generator of Fig. 208 had been connected to line wires that were already joined between some other source and load, the prime mover could still cause the machine to generate power into the line wires. However, if the mechanical power of the prime mover were shut down, the shunt machine would still continue to run, as a *motor,* and in the same direction of rotation as the prime mover had been driving it. Review the equivalent circuit of Fig. 209 for the generator action. The equivalent circuit for motor action is shown in Fig. 217. Note that in the motor the armature current enters from the line to

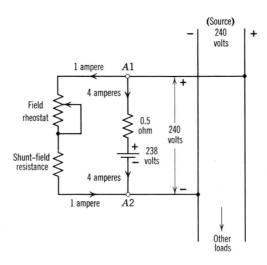

Fig. 217. Current directions when the machine of Fig. 208 is operated as a motor.

11: D-C Machines   249

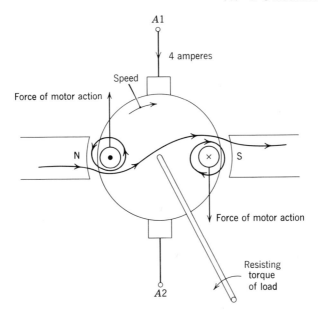

Fig. 218. Torque and force directions in a motor.

charge the internal voltage as if it were a storage battery. The product of the internal voltage and this charging armature current represents the power, in watts, that is being converted from electrical form into mechanical form on the shaft of the motor. The sketch of Fig. 218 shows the pertinent directions of the forces of motor action and the resisting forces of a mechanical load. Figure 216 should be inspected for comparison of the force directions when a machine is generating. Since the rotor conductors are spinning through the same magnetic flux at essentially the same speed and in the same direction as in the generator, the shunt motor has essentially the same internal voltage, as indicated in Fig. 217. (Because of the direction of armature current, the internal voltage in the generator is slightly greater than the line voltage. The internal voltage in the motor is slightly less than the line voltage.) Therefore, it is clear that there are *both* a force of motor action *and* an internal generated voltage in a *motor*. It is this inherent presence of both basic principles that makes a motor essentially reversible for use as a generator, and vice versa.

In a motor, the internal generated voltage is often referred to as the *back voltage* of the armature, or as the *counterelectromotive force*. Although there is in principle no difference between the Thevenin voltage of a generator and the Thevenin voltage of a motor, the use of the phrase, back voltage, helps to keep the polarity fixed in mind. As noted above, the back voltage in a motor is just a little smaller than the line voltage, the difference being the drop in the internal armature resistance.

A shunt motor has an inherent hazard associated with a greatly weakened magnetic field. Assume that some accident interrupts the field circuit of the machine of Fig. 217 while the machine is motoring. Assume that the magnetic flux of the machine drops suddenly to a residual value only 10% of the operating value. If, for a moment, the speed stays the same, Faraday's law shows that the internal voltage would now be only about 10% of the operating internal voltage, or now about 24 volts. Kirchhoff's voltage law then shows that there would be $240 - 24 = 216$ volts across the 0.5-ohm equivalent internal resistance of the armature circuit. From Ohm's law applied to this resistance, we can see that the armature current would have a momentary value equal to $216/0.5 = 432$ amperes. In some machines, the commutator may not be able to switch this magnitude of current successfully. If an arc forms on the commutator, the arc may progress from one brush set to the next of opposite polarity in a vivid display known as *flashover*. It is customary to choose fuses or other protective apparatus such that this dangerous condition may not occur.

However, there is another danger with the shunt machine with greatly weakened field. Although the magnetic flux has been reduced by a factor of 10 in the example given above, the armature current has been *increased* by a factor of 86. Therefore, the effect of the armature current *charging* the internal voltage shows an increase in internal power of $\frac{86}{10}$, or almost nine times the value previously existing. Since this figure represents power converted from electrical to mechanical form, and since the speed is momentarily the same, this calculation shows that the mechanical torque of the motor suddenly increases by a factor of about 9 when this open-field accident occurs. (In practical machines, the changes may not be sudden or quite as drastic as indicated here. All d-c machines have certain time-rate-controlling properties that have been ignored in the equivalent circuit.) This increase in torque causes the machine to speed up rapidly, with the armature current coming back down to more usual values as a result. If the armature current should get back down to 5 amperes, as in Fig. 217, the internal voltage would then be 238 volts once more. *But,* Faraday's law then shows that the speed would have to be ten times as great as the normal condition: with the same armature conductors turning through only 10% of the normal flux, the speed of motion must be ten times as great to produce the same voltage. For most designs of d-c machine, the rotor parts cannot withstand the resulting centrifugal forces, especially the copper parts. The overspeeding machine literally throws itself apart.

At first appearance, it may seem wrong that making the magnetic field of a shunt motor *weaker* causes the machine to speed up. However, it is the closeness of size that the internal voltage has with respect to the line voltage that controls this behavior. In summary, if the shunt field of a motor is greatly weakened or the circuit of the field is opened, the motor will probably overspeed destructively, or suffer flashover; or both disasters may occur.

17. Starting Resistance. For starting any motor other than one of small size or special design, the circuit must not apply full line voltage to the armature suddenly. Recall from Section 11-2 that the internal generated voltage (in accordance with Faraday's law) of a machine depends on the speed of the machine. When a motor is standing still, its speed is zero; therefore, its internal generated voltage is zero, even if there is a magnetic field set up by the field windings. If the full line voltage is applied to the armature terminals with the machine standing still, the resulting current is determined from Ohm's law, with the full line voltage applied directly to the small armature-circuit resistance of the machine. Because the usual d-c machine has a very low internal armature resistance, the starting current in the armature would be excessive, either burning the insulation on the conductors or causing a failure in commutation as soon as the armature begins to turn.

Therefore it is necessary that a *starting box* be provided for the motor, the box containing an adjustable resistor for insertion in series with the armature circuit of the machine. This resistor, often called a *starting resistance*, adds its resistance to the internal resistance of the armature to determine a reasonable magnitude of current when the full line voltage is applied to the series combination of the armature and this starting resistance.

In the previous section we noted that the internal generated voltage of a motor is often called the *back voltage* of the armature. With the aid of Kirchhoff's and Ohm's laws, we may state the relation for the current as follows:

$$\text{current (in the armature)} = \frac{\text{line volts} - \text{back volts}}{\text{resistance of armature circuit}}$$

With the motor standing still, the back voltage is zero. Thus, the armature will have a safe current through it only if the total resistance of the armature circuit is increased by the proper amount, added in series in the starting box.

252 Essentials of Electricity

Example 1. The resistance of the armature of a 2-horsepower motor is 0.4 ohm. What current will be set up in the armature if it is connected directly to a 115-volt line at standstill?

$$\text{amperes (in armature)} = \frac{\text{volts (across armature)}}{\text{ohms (of armature)}}$$

$$= \frac{115}{0.4} = 288 \text{ amperes}$$

Since the normal current for such a motor is about 16 amperes, this inrush current of 288 amperes would be disastrous for the motor.

Example 2. When the motor in Example 1 is running near its normal speed, the armature develops a back voltage of 109 volts. What is the current in the armature when the full-line potential of 115 volts is directly connected to the armature terminals?

$$\text{amperes (in armature)} = \frac{\text{line volts} - \text{back volts (within armature)}}{\text{ohms of armature circuit}}$$

$$= \frac{115 - 109}{0.4} = \frac{6}{0.4} = 15 \text{ amperes}$$

Example 3. If the normal current for the motor in Example 1 is taken to be 16 amperes, how much resistance must be provided in the starting box to limit the starting current to 1.5 times the normal current?

$$\text{starting current} = 1.5 \times 16 = 24 \text{ amperes}$$

With no back voltage at standstill, the *total* armature-circuit resistance determines the current:

$$\text{total resistance} = \frac{\text{line volts (across entire circuit)}}{\text{amperes (in series circuit)}}$$

$$= \frac{115}{24} = 4.79 \text{ ohms}$$

Because the armature internal resistance is 0.4 ohm, the resistor in the starting box must add the difference:

$$\text{starting resistance} = 4.79 - 0.4 = 4.39 \text{ ohms}$$

Figure 219 shows a schematic diagram of a shunt motor connected to an adjustable starting resistor *SR*. When the line switch is closed into its ON position and when arm *C* is placed on the first point (or tap) of the starting resistor, the full-line voltage is applied directly to the field winding of the shunt motor. Therefore, the field current for the motor is established quickly at its normal full value. Current is also set up in the armature circuit, i.e., through the armature in series with the entire starting resistance. The size of the total starting resistance has been chosen to limit the armature current at standstill to a safe value.

With full-field current, the normal strength of magnetic flux is set up in the iron of the motor. The safe value of armature current creates

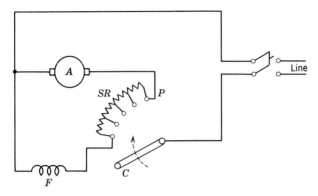

Fig. 219. Essential connections of a shunt machine for operation as a motor with a starting resistance.

the forces known as motor action, which cause the armature to accelerate to some low speed. When the armature conductors are turning at any speed through the magnetic flux lines of the machine, a back voltage is created in accordance with Faraday's law. At the low speed reached by the armature when arm C is on the first point, the small back voltage of the armature opposes the line voltage, bringing about a reduction in the armature current below the starting value. Therefore, arm C may then be moved to the next tap point of the starting resistance.

At the second point, the first section of starting resistance has been cut out of the armature circuit by arm C. Although the remaining resistance in the armature circuit is now smaller than at standstill, the net voltage difference between the full line voltage and the internal generated back voltage is also smaller. As a result, although the armature current increases suddenly when arm C is moved to the second point, the peak of current reached will still be a safe value if the tap point has been properly selected for the particular motor and its attached load.

The increase in armature current will cause another increase in torque due to the forces of motor action, and the armature will accelerate to a higher speed. Arm C may be gradually moved up as the speed of the motor increases, with the arm finally resting at point P for normal operation. At point P, the only resistance remaining in the armature circuit is the internal resistance of the armature itself. All of the starting resistance has been removed from the armature circuit at point P.

Note that the field current also passes through arm C and through the sections of starting resistance *below* the position of arm C. It might seem that this resistance added to the field circuit would cause the field of the motor to be weakened, perhaps dangerously. However, a consideration of the sizes of the resistances in the field circuit shows that the field is hardly weakened at all. The resistance of the field winding itself for the motor of Example 1 might be 250 ohms. In Example 3, we found that the *total* starting resistance needed for the *armature circuit* was only 4.39 ohms. Therefore, at point P the field-circuit resistance has been increased to 254.4 ohms from the value of 250 ohms when arm C was at the first point. This very small *increase* in field-circuit resistance will cause a small enough *decrease* in field current that the change may be ignored. An obvious advantage of such a connection is that only one movable contact is involved in the starting apparatus. If arm C should happen to break contact with the field circuit, the armature circuit is ruptured simultaneously. In this way the overspeed danger of the shunt motor is averted for this type of accident.

Prob. 21-11. In Fig. 219 the size of starting resistance SR is 4.7 ohms, and the internal resistance of the armature itself is 0.8 ohm. What current is set up in the armature at standstill when the motor is started from a 115-volt line?

Prob. 22-11. Field winding F of Fig. 219 has a resistance of 230 ohms. When arm C is on the first point of the starting resistance with a 115-volt line potential, what is the size of the field current?

Prob. 23-11. What is the current in the line wires for the motor of Probs. 21-11 and 22-11 just after arm C is set at the first point of the starting resistor?

Prob. 24-11. What is the field current in the motor of Probs. 21-11 and 22-11 when arm C is at the running point, P?

Prob. 25-11. When the motor of Prob. 23-11 is driving its normal load with all of the starting resistance cut out of its armature circuit, its back voltage is 106 volts. What is the size of the armature current? What current exists in the line wires for this loaded condition?

Prob. 26-11. The armature resistance of a 4-horsepower 230-volt motor is 1.5 ohms. Its field resistance is 450 ohms. When the motor is loaded to its full rated horsepower, its line current is 16.5 amperes. What starting resistance is necessary if the starting current must not exceed 1.5 times the full-load current?

Prob. 27-11. For the full-load condition, what is the back voltage of the motor of Prob. 26-11 when all of the starting resistance has been cut out of the armature circuit?

18. Manual Starting Boxes. A simple scheme for starting a shunt motor is to provide a main switch (sometimes called a *disconnect*) for the line wires, fuses, and a *manual starting box*. The starting box contains the tapped starting resistance, a holding electromagnet that retains the movable arm in the running position, and certain additional safety devices.

For preventing the overspeeding destruction of a shunt motor when the field circuit is accidentally opened, a device known as a *no-field release* is commonly employed. The movable arm of the starting resist-

11: D-C Machines

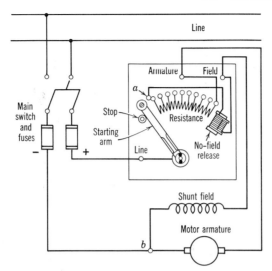

Fig. 220. Diagram of a 3-point starting box showing the no-field release.

ance has a strong spring attached, which tends to return the arm to a disconnected position below the first starting point. The person who operates the starting box must work the arm gradually over the starting points, against the force of the spring. When the movable arm reaches the final running tap point, an iron sleeve on the arm comes in contact with the poles of an electromagnet. This construction is shown schematically in Fig. 220, where the appropriate electrical connections are also designated. Since the field current of the motor must pass through the winding of the holding electromagnet, the arm may be released by the operator when the running point is reached. Because the iron sleeve on the movable arm provides a path of high permeability for the flux lines of the electromagnet, a strong holding force is developed when the arm is in the running position. If, however, the field circuit should become open accidentally, the current for the holding magnet is also cut off. The force of the spring will return the arm to the OFF position against a stop, so that all power will be removed from the motor. In this way the overspeed danger of the shunt motor with open field is avoided.

Figure 221 shows a commercial form of a starting box with no-field release. Note that there are just three terminals on the box for external connections: one to *line,* one to *field,* and one to *armature.* Such a box is known as a *three-point box;* since the coil of the electromagnet must

256 Essentials of Electricity

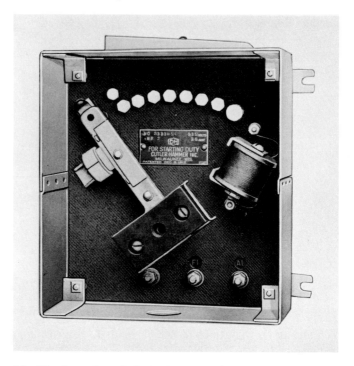

Fig. 221. Internal panel of a 3-point starting box. *Cutler-Hammer Inc.*

be wired in series with the field winding of the motor, it is important that the correct coil be chosen in relation to the size of the field current of the motor. In order to prevent contact with the live parts inside the starting box, a *dead-front* construction is common. Figure 222 shows the covering panel for a three-point box. When the panel is in place, the external arm attaches with an insulated grip to the internal movable arm illustrated in Fig. 221.

Let us trace out in detail the connections in Fig. 220. With the positive terminal of the line connected to the *Line* terminal as shown, the line current for the motor enters the box and is carried along a conductor on the movable arm. The tip of the arm becomes the dividing point where the line current separates into its two components of field and armature current. If the arm is at point *a*, the field-current component goes through the electromagnet of the *no-field release*, out through the terminal marked *Field,* through the shunt field of the motor back to the negative side of the line. The rest of the line current available at the tip of the movable arm is the armature current. It threads through all of the starting resistance, out of the box at the terminal marked *Armature,* through the armature and back to the negative side of the line. When the movable arm is in the running position up against the magnet of the no-field release, the starting resistance has been cut out of the armature circuit for normal operation of the motor. In the running position, the field current of the motor passes through all of the starting resistance of the starting box. As noted in the preceding section, this small extra resistance in the field circuit makes very little difference in the size of the field current of the motor.

Note how the motor, the starting box, and the line connections are arranged. The terminal on the box marked *Line* is connected to one of the line wires, often the positive line. The terminal marked *Field* on the box is connected to one of the field terminals of the motor. The terminal marked *Armature* on the box is connected to one of the armature terminals of the motor. Thus we have one end of the line joined through the box to *one* end of the field and to *one* end of the armature. To complete the basic parallel connection for the shunt motor, we connect the *other* end of the field to the *other* end of the armature together

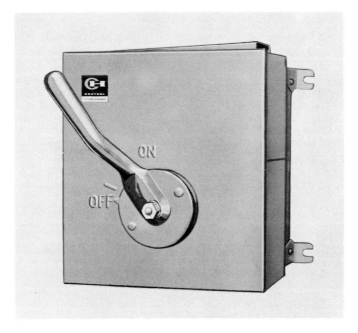

Fig. 222. Dead-front construction for a starting box. *Cutler-Hammer Inc.*

258 Essentials of Electricity

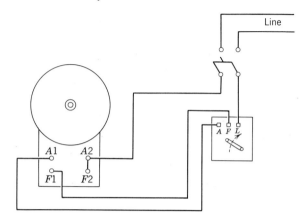

Fig. 223. External connections of a shunt motor to line wires through a 3-point starting box.

and join this common point back to the *other* end of the line, usually the negative line.

Figure 223 shows a wiring diagram to illustrate these connections. Point L on the box is joined directly to one side of the line switch; point F is joined to $F1$ (field terminal); point A is joined to $A1$ (armature terminal). Then $A2$ and $F2$ (the other field and armature terminals) are joined together and brought directly to the other side of the line switch.

Prob. 28-11. With the outline sketches in Fig. 224, show how the shunt motor should be connected through the starting box to the line.

Prob. 29-11. Draw the inside connections of the box sketched in Prob. 28-11.

Prob. 30-11. Draw the inside connections for the box sketched in Fig. 223.

Another form of the manual starting box is also common. A different

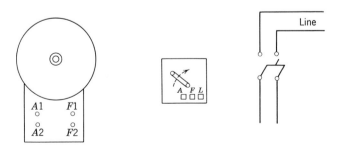

Fig. 224. Shunt motor, starting box, and line.

design of the coil for the holding electromagnet permits the coil to be connected directly across the line wires, instead of being in series with the field as in the three-point box. Because this shunted arrangement requires another connection to the box, such a starting box is called a *four-point box*. Its circuitry releases the movable arm if the line voltage should fail, thereby stopping the motor. If the circuit is not shut down when the line voltage fails, it is possible that full-line voltage may be reapplied to the armature while it is at standstill, with the resulting damage noted in the preceding sections. The holding mechanism of a four-point box is therefore known as a *no-voltage release*. Of course, the three-point box also performs this same function, but the coil of the electromagnet must be chosen for the motor with which the box is to be used.

Figure 225 shows a schematic diagram of a four-point box and its connections. Note that there are two *Line* terminals on the box. Internally, there is a direct connection from the first tap point A, on the starting resistance to the *Field* terminal of the box. There is another path from point A through the electromagnet of the no-voltage release to one of the *Line* terminals. Otherwise, the arrangement of a four-point box is the same as that of a three-point box.

For connecting a shunt motor through a four-point box to line wires, observe the following steps:

CONNECT:

one side of the line to one box terminal marked *Line*
one side of the field to the box terminal marked *Field*
one side of the armature to the box terminal marked *Armature*
other side of the line to the *other* terminal marked *Line*

NOTE:

The *Line* terminal connected to the movable arm should be the first connection made. (On a box where the internal connection is not indicated, join line wires *first* to line terminals B and C. A voltmeter placed across points C and D will read zero if D is internally connected to C as shown; between B and C the meter will indicate the full line voltage. The zero and full-voltage indications will be reversed if the other *Line* is connected to the movable arm. An ohmmeter may also be used to find the correct terminals, if all power is removed from the starting box.) The other ends of the field and armature windings are to be connected together and to the side of the line *not* connected to the movable arm.

Whenever a shunt motor must be reversed, it is necessary to reconnect its circuits so that torque is produced in the opposite direction.

260 Essentials of Electricity

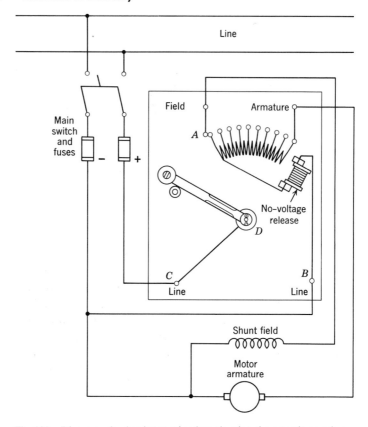

Fig. 225. Diagram of a 4-point starting box showing the no-voltage release.

The torque will be reversed if *either* the armature current *or* the field current is reversed. If *both* currents are reversed, the torque will be in the original sense, and the motor will run in the original sense.

For compound motors, especial care must be taken in reversing the direction of operation, so that the *kind* of compounding will not be altered in the process of reversal.

Prob. 31-11. With the outline sketches of Fig. 226, show the connections of the shunt motor with the four-point starting box and the line wires.

Prob. 32-11. Show the internal connections of the box outlined in Prob. 31-11.

19. Automatic Starters. Motors are often operated with the aid of devices employing magnetic contactors and relays. A push-button station located conveniently for the operator initiates START and STOP actions. The magnetic contactors internal to the associated automatic starting

11: D-C Machines   261

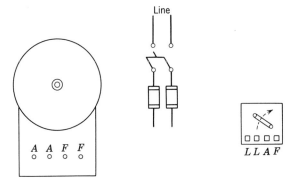

Fig. 226. Shunt motor, line, and 4-point starting box.

device take over the function of inserting the correct starting resistance in series with the armature of the motor, thereafter cutting the resistance out in steps at the appropriate time. Most automatic starters also have protective overload devices built in, which will shut down all equipment when the motor is overloaded seriously. Figure 227 shows a commercial reduced-voltage starter having just two steps of starting resistance. The resistors that are inserted into the armature circuit are shown in a separate part of the enclosure at the left of the box. A typical wiring diagram for such a starter, with its motor and associated push-button station, is shown in Fig. 228. The operation of this circuit will be traced through carefully.

When all of the power has been removed from the circuit, all relays and contactors return to their *normal* condition. It is in this *normal* condition that all the contacts are shown in the diagram. With all power removed, the motor will eventually come to rest.

With the line disconnect switches turned ON, depressing the START button establishes a circuit through the normally closed OL contacts, the normally closed STOP button, the START button, and the main contactor coil, M. The current through the coil M operates the main contactor, which closes the two series-connected contacts, marked M, in the armature circuit. This action applies line voltage to the following circuit: through contacts M, the heating element for overload device OL, the total starting resistance, and the armature of the motor. In a parallel path beginning at the starting resistance (as in a manual starting box), the shunt field is also connected essentially to the full line voltage. Therefore, depressing the START button causes the correct connections to be made for starting the motor.

The holding contact M in parallel with the START button also closes

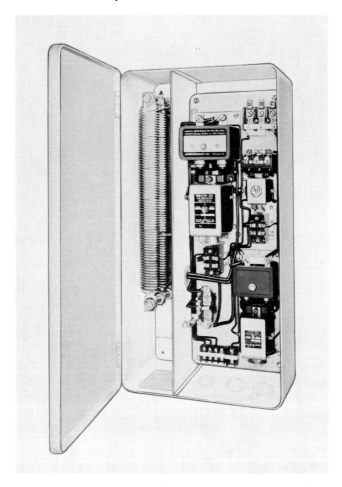

Fig. 227. Nonreversing d-c reduced-voltage starter for constant-speed motors. *Allen-Bradley Company.*

at the same time as the main motor contacts, thereby sealing main contactor M onto the line, as described in Chapter 10. A pneumatic timer associated with contactor M allows a short time to elapse while the motor is accelerating. When this preset time has elapsed, the timer allows the timed contact $M(T)$ to close in series with contactor coil $1A$. This contactor is known as the *first accelerating contactor* for the automatic starter. When the time-delay contact $M(T)$ has closed, contactor $1A$ closes its heavy-current contact $1A$, shorting out the portion of the total starting resistance connected from the OL terminal to the first tap

11: D-C Machines

point. This action allows the motor to accelerate further with the reduced starting resistance present in the armature circuit.

Another pneumatic timer is attached to contactor 1A. It begins timing action when heavy-current contact 1A closes. After a short time has elapsed, the timer allows time-delay contact 1A(T) to close, connecting the line power to the coil of the second accelerating contactor 2A. Operation of this contactor closes heavy-current contact 2A, shorting out the remainder of the starting resistance. The motor then accelerates to its final operating speed without any additional armature resistance in the circuit.

When the STOP button is depressed momentarily, power is removed from the coil of main contactor M. Its heavy-current contacts return to the normally open position, disconnecting all power from the motor. Power is likewise disconnected from the coils of the accelerating contactors. When their contacts open again, the total starting resistance is connected in series with the armature of the motor, ready for another start. Since the sealing contact, M, on the main contactor also opens, the STOP button needs to be depressed only momentarily; all circuits return to their original condition without further attention from the operator of the equipment.

Sometimes the armature current may become excessive. If the work

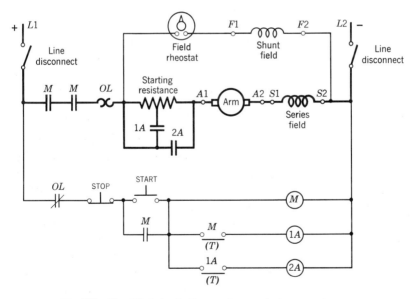

Fig. 228. Simplified circuit diagram for a typical automatic starter.

load is increased beyond the capacity of the motor, the large armature current that results may overheat and damage the motor. During the starting operation, the motor may be stalled by some accident to its connected load. Whatever the reason for the excessive armature current, the automatic starter will shut down the equipment through operation of its overload device. The excessive armature current will cause heating in the OL element in series with the main heavy-current contacts, M. The heater element bends or melts, allowing a spring action to open the normally closed OL contacts. Since these contacts are in series with the STOP button, their opening shuts down the equipment in exactly the same way as the STOP button does. In most automatic reduced-voltage starters, the operator must depress a *reset* button on the starter box to reclose the overload contacts after the overloading cause has been removed.

20. Series Motors. As noted above for generator operation, a field winding may be constructed for a d-c machine with large enough wire and a sufficient number of turns that this field winding may be placed in *series* with the armature. Although the series generator is somewhat limited in its application, there are important uses for series motors. For a shunt motor, the speed tends to be reasonably constant from no load to full load, for a fixed field current. For a series motor, the torque is high for slow speeds and low for high speeds. Thus, without the benefit of gear shifting or other torque-converting schemes, the series motor can accelerate heavy loads from standstill to full speed with relatively simply control schemes. For driving subway cars or other railway vehicles, this torque-speed relationship for the series motor matches very well the action needed for fast acceleration of a load on iron wheels driven over iron rails. Since a series motor when unloaded acts like a shunt motor with its field opened (i.e., the unloaded series motor tends to run away), such a motor is never used unless permanently geared, or directly connected, to its load.

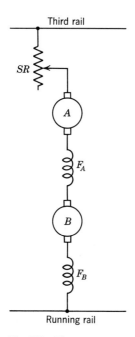

Fig. 229. The two motors are in series with each other and with part of starting resistance SR.

Traction motors for rapid-transit vehicles are basically series motors, even if certain designs include special compound connections for use in stopping the vehicle. Both multiple-unit cars and locomotives often have *even numbers* of series motors geared to the driving wheels. This arrangement permits a starting scheme known as *series-parallel control.* Figures 229 and 230 outline the essential parts of this scheme for a vehicle having two traction motors. When the car is at standstill, the two motors are started in series, as indicated in Fig. 229. The starting resistance is provided by a bank of high-current-capacity resistors, usually mounted under the vehicle. As the speed of the rapid-transit car increases, manual or automatic control gradually shorts out the starting

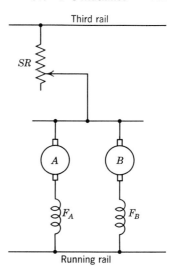

Fig. 230. The two motors are paralleled to make a motor grouping that is in series with part of starting resistance SR.

resistance until the two motors in series are directly connected to the full system voltage. Therefore, each motor has one-half of the line voltage applied. Then the control scheme makes a transition in the connections to the parallel arrangement shown in Fig. 230, with the full system voltage. Therefore, each motor has one-half of the line accompanies the changeover. Further operation of the devices that short out the starting resistors finally brings about full-speed operation of the vehicle with each series motor operating from the full system voltage.

The scheme of putting the two motors in series at standstill allows the car to be started on half the current it would require if the motors were started in parallel. More power would be wasted in the starting resistors each time the car started. Since frequent starts are necessary for an effective rapid-transit schedule, the power loss in the cars and the current levels in the third-rail or trolley system become factors of prime importance. The series-parallel scheme of starting two series motors essentially makes each motor act as the starting resistance for the other, while both motors develop torque for acceleration.

21. Some Precautions for Operating Shunt and Series Motors. A shunt motor *races* when the field circuit is broken, if the armature circuit is not also broken. Therefore:

Never open the field of a shunt motor.

A series motor *races* when it has no load connected to it. Therefore:

Never start an unloaded series motor and never remove all the load from a series motor while it is running.

In a manual starting box destructive arcing occurs between the first tap point and the contact on the movable arm, if the arm is moved backward. Stop a motor by opening the main switch when possible.

Never pull back the arm of a starting box.

For starting use only, the resistors in a manual box are not designed to carry the heavy armature current continually. The resistors will burn up if the arm is left at any of the intermediate tap points.

Never allow the movable arm of a manual starting box to remain at an intermediate tap point more than a few seconds.

NOTE: Boxes or controllers intended for speed control (and thus continuous operation at an intermediate point) are always clearly marked for this type of operation.

22. Some Basic Machine Troubles. It is the purpose of the following paragraphs to discuss briefly some methods for locating and correcting troubles which appear in motors and generators after they have been installed. Such tabulation is intended as a guide for students who will use motors and generators. Only simple tests that can be made with an ammeter and voltmeter have been indicated.

In particular it should be noted that these instructions apply to shunt or compound machines operating only in the range of 115 to 230 volts. For work around machines of higher voltage, much more care must be exercised than has been advised here to prevent hazardous electric shocks.

Certain happenings are *signs* of trouble, the *causes* for which are perhaps hidden. An expert has the skill to read these signs and go fairly directly to the cause of the trouble. Usually the cause is easy to remedy once it has been located. The greater difficulty is in the diagnosis, or recognition of the cause of the trouble from the signs of trouble. Consequently, in the material that follows more emphasis is placed on the diagnosis than on the remedy. In any event, the manufacturers' instructions for their own equipment should be consulted, if possible, for authoritative maintenance and repair procedures.

D-C Machine Troubles

SIGNS	CAUSES
1. Sparking at Brushes (Section 11-23)	1. Overload 2. Weak field 3. Poor brush contact 4. Commutator rough or off center 5. Armature winding broken or short-circuited 6. Brushes set wrong
2. Noise (Section 11-24)	1. Rattle—loose parts 2. Bumping—too little end play 3. Squeaking—dry brushes 4. Rubbing and pounding—armature hitting pole 5. Excessive vibration—unbalanced armature
3. Hot Bearings (Section 11-25)	1. Lack of lubricant 2. Grit 3. Not enough end play 4. Bearing too tight 5. Poor alignment 6. Crooked shaft 7. Hot commutator 8. Rough shaft
4. Hot Field Coils (Section 11-26)	1. Too large field current 2. Moisture in windings
5. Hot Commutator (Section 11-27)	1. Near some hotter part of machine 2. Sparking under brush 3. Poor brush contact
6. Hot Armature Coils (Section 11-28)	1. Overload 2. Damp windings 3. Short-circuited coils
7. Fails to Build Up (Section 11-29)	1. Field connections reversed 2. Wrong direction of rotation 3. Speed too low 4. Field circuit open 5. Not enough residual magnetism 6. Machine short-circuited
8. Too Low Voltage (Section 11-30)	1. Too much resistance in field 2. Overload 3. Speed too low 4. Some poles reversed or short-circuited
9. Too High Voltage (Section 11-31)	1. Too strong field 2. Speed too fast

268   Essentials of Electricity

10. Motor Fails to Start (Section 11-32)	1. Wrong connections 2. Open circuits in connecting wires 3. Field weak 4. Overload 5. Friction excessive
11. Too High Speed (Section 11-33)	1. Too much field-rheostat resistance 2. Connections wrong 3. Open field circuit
12. Too Low Speed (Section 11-34)	1. Overload 2. Too little field resistance 3. Excessive friction 4. Short circuit inside armature

23. Sparking at Brushes. Any or all of the following causes may bring on sparking:

Test for and correct as needed:	1. Overload 2. Weak field 3. Poor brush contact 4. Commutator rough or off center 5. Armature winding broken or short-circuited 6. Brushes set wrong

(1) To Test for Overload:

FIRST: If the machine is a *generator,* note the reading of line ammeter. If the current indicated is above the rating of the machine:

Make a rough estimate of the current taken by all the electrical loads attached to the line wires from the generator. If the estimate agrees with the reading of the ammeter, the only remedy is to reduce the number of loads attached to the generator.

SECOND: If both line wires of the generator were intended to be ungrounded, and if the estimate is considerably less than the actual ammeter reading, test for a ground: Attach one terminal of a lamp or of a voltmeter to one line wire and the other terminal to a connection to ground (earth), such as a water pipe. If the lamp glows brightly, or if the voltmeter reads the voltage of the line, either sign indicates that the *other* line wire has a direct connection to ground somewhere. Find this unwanted ground and repair it so that the voltmeter does not read when connected between either wire and ground.

If the voltmeter reads zero in the first test (indicating no unwanted ground), there may be unwanted connections or leaks from one wire to the other. Such leaks may occur also in a normally grounded system.

Careful checking over all wiring connections will help to locate such faults. These unwanted paths for current may often be detected by the heating they cause in the connection or wiring device where they occur.

THIRD: If the ammeter reading is not above the capacity of the machine, there may be unwanted leaks or grounds *near* or *at* the generator, between the machine and the point in the circuit where the ammeter has been connected.

If the machine is a *motor,* the line ammeter will give a clear indication of overload. If the overload is in the machinery driven by the motor, and if the motor is belted to its load, there will be excessive tension in the tight side of the belt, possibly accompanied by slipping and squeaking.

FOURTH: Stop the machine and feel the armature coils. If they are *all* uncomfortably warm to the touch, this heating is a sure sign of overload; however, there can be an overload that will cause sparking without overheating the armature coils.

If the overload is brought about by friction in a motor, the line ammeter will show a large current even when its mechanical load is uncoupled. This "no-load" current should not be more than about 8 percent of the full-load current. For some help in eliminating such friction, see Section 11-24.

(2) To Test for Weak Field:
FIRST: The speed will be excessive if the machine is a motor. The sparking will be worse when the motor is starting. A weak field may very likely be due to wrong connections. The sense of the poles may be checked with a compass; a separate check should be made with only the shunt field excited, and then another check should be made for the series field, if present. The sense of the poles should be alternately north and south around the frame of the machine.

SECOND: Broken circuit in a field coil (opens the circuit for all the coils).

Test a motor by applying power only to the field circuit. If opening this connection shows no spark as the circuit is ruptured, or if an ammeter shows no field current with the power ON, the circuit has a break in it. Each coil should then be checked separately in the same way: it may be necessary to apply reduced voltage (from dry cells or other source) to avoid damage to the individual coils. By this means, or by any other appropriate continuity checking scheme, the open-circuited coil or coils may be located.

THIRD: There may be a short circuit in one field coil (weakening the field of one pole only).

Hold a piece of iron (such as a screwdriver) near one pole, and then each of the others in turn. If one pole produces a decidedly weaker magnetic pull on the iron than the others, its coil probably is short-circuited. With all the coils excited, a low voltmeter reading across any one coil also indicates a short-circuited coil. For this defect, rewinding the coil is usually necessary.

(3) To Test for Poor Brush Contact:
FIRST: Note the appearance of the commutator. It should have a clean, smooth chocolate-brown color.

SECOND: See that the brushes bear evenly over all their bearing surfaces. Any brushes which do not press down evenly should be ground with sandpaper until they fit the curvature of the commutator.

THIRD: Press each sparking brush separately. Note how it fits in its holder. Adjust the tension of the spring, noting whether tightening or loosening diminishes sparking.

(4) To Test for Rough Commutator: Determine whether the mica insulating strips project above the copper commutator segments, or if one of the segments has worked up higher than the others. Note rough spots caused by melting due to momentary overloads. Sandpaper held on a wooden block that has been curved to fit the commutator will remedy these defects. *Never use emery on a commutator.*

NOTE: If grooves have been worn in the commutator, or if the commutator has become so much off center that the brushes ride up and down as the commutator revolves, the commutator should be turned down, in a lathe or by other appropriate tools. Sufficient end play will help prevent grooves.

(5) To Test for Faulty Armature Coils:
FIRST: Short circuits in armature coils can usually be located by noting that some of the windings are very warm after the machine has been operated. A short-circuited coil should usually be suspected as the cause of such heating, especially if the sparking occurs at only one point on the commutator. For further verification, apply a low voltage to the armature circuit only. Then connect a low-range voltmeter between adjacent commutator segments, observing the reading for each adjacent pair around the circumference. The shorted coil is connected between the segments having no reading, or a very low one.

SECOND: If the sparking is due to an open-circuited coil, the sparks will be vicious and will always occur at a fixed spot on the commutator. A low voltage arranged in series with an ammeter should be applied to adjacent commutator segments, a markedly lower reading of the ammeter indicating the open-circuited winding. Of course, any other suitable continuity checker may be employed.

For broken, shorted, or grounded coils, repair usually involves replacement or rewinding of the faulty coils in the armature.

(6) To Test Setting of Brushes: Machines with interpoles have their brushes set in the proper position by the manufacturer. Ordinarily, the brush carriage should be adjusted only if there is reason to believe that the factory setting has been disturbed. Look for a scratch or other mark put on by the manufacturer for the proper position.

24. Noise. All machines hum and vibrate a little; however, there is something wrong with the machine when any unusual noises occur, such as those listed below. These should be investigated and corrected.

1. Rattle		1. Loose parts
2. Bumping		2. Too little end play
3. Squeaking	means	3. Brush trouble
4. Rubbing and pounding		4. Armature hits pole
5. Excessive vibration		5. Poor alignment or unbalance

(1) Rattle: Look the machine over for loose nuts or other parts and tighten them.

(2) Bumping Against Bearings:

FIRST: Note whether or not the collar strikes the bearing as the armature shaft travels back and forth along its axis. If this is the cause of the bumping, stop the machine and set the collar to allow more end play.

SECOND: If the machine is direct-coupled to another, the pounding is probably due to poor alignment of the machines. With the machines running, hold a pencil firmly fixed and gradually bring it near a smooth place on the coupling until the pencil just touches. Then the machines are stopped. If the mark made by the pencil does not extend all around the coupling, the coupling *bulges* on the side where the pencil touched. The machine should be realigned to force this point inward a little.

(3) Squeaking: One or more of the brushes may be the source of this noise.

FIRST: Being careful not to open the armature circuit, try lifting each brush gently in turn. Find the brushes which make the most noise, and readjust the tension; be sure the brush holder allows for proper play.

SECOND: For machines intended to rotate in only one direction, the brush holders may slant back slightly to avoid a sharp angle at the leading edge of the brush. Therefore, be sure that the direction of rotation is correct, and that the brushes fit the curvature of the commutator correctly. This curvature can be obtained by holding a strip of sandpaper firmly on the commutator while rocking the armature back and forth, letting the sandpaper wear down the brushes. On large machines, pull the sandpaper between the brush and armature, holding the brush firmly against the paper. Although newly replaced brushes usually squeak at first, this noise should disappear after a run-in period of several hours.

(4) Rubbing and Pounding: This noise may result if the armature rubs on one of the pole faces. Stop the machine to examine the pole faces and the surface of the armature. If the armature winding has worked loose anywhere, fasten it in place again. If the bearings are worn or misadjusted so that the armature does not lie in the center of the stator, repair or replace the bearings and adjust them for proper clearance under all pole faces.

(5) Excessive Vibration. Feel the vibrations by placing your hand on the frame of the machine. If possible, change the speed of the machine (providing other running conditions are not disturbed greatly). If the machine still vibrates badly, try changing the alignment of the bearings, one at a time. As a last resort, take out the armature and balance it on two knife edges, as shown in Fig. 231. Roll it gently, setting it several times with different parts uppermost. If it always tends to come to rest with the same part down, the armature or coupling is not properly balanced. The addition of nuts or other small weights to the light side may afford a satisfactory remedy. Otherwise, expert help in achieving dynamic balance may be needed.

25. Hot Bearings. If there is a smell of burning oil, or if the bearings are too hot for the hand to be held on them, any of the following causes may be present:

Test for and correct in this order:
1. Lack of lubricant
2. Grit
3. Not enough end play
4. Bearing too tight
5. Poor alignment
6. Crooked shaft
7. Hot commutator
8. Rough shaft

(1) Lack of Lubricant: Check that the bearings are adequately greased or supplied with oil. If oil cups are the source of oil, check that they are full and delivering oil to the bearing surfaces. Observe whether oil rings, if any, rotate freely and bring up oil to the bearing. Expert advice from the machine manufacturer or from a lubrication specialist

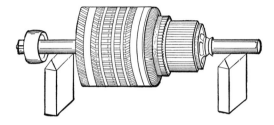

Fig. 231. Testing the balance of an armature.

should be sought if there is doubt about the kind of oil or grease to be used.

(2) Grit in Oil: If a little oil from the bearing feels gritty when rubbed between the fingers, the machine should be stopped; the shaft and bearing should be cleaned thoroughly, and the old oil should be replaced with clean oil.

(3) Not Enough End Play: If the shaft collar keeps bumping against the bearing, there is not enough end play allowed for the iron of the armature to "float" properly in the pull of the magnetic-field lines. Stop the machine and set the collar further from the bearing.

(4) Bearing Too Tight: Adjust the cap of the bearing for a looser fit; notice if the bearing runs cooler.

(5) Poor Alignment: When one machine is direct-connected to another, the alignment of the two shafts may be poor. Using the test outlined in Section 11-24 under *Bumping,* note whether or not the coupling runs true. If not, make a new alignment and test until the coupling does run true.

(6) Crooked Shaft: Observe the armature for evidence of wobbling, an indication that the shaft is crooked. With the machine stopped, turn the armature by hand while watching carefully for wobble. If this trouble is the cause of the heating, the remedy is a new shaft or complete rotor structure.

(7) Hot Commutator: Stop the machine and feel of the commutator and bearing. If the commutator is evidently hotter than the bearing, the heat probably comes from the commutator. This trouble can be eliminated in the manner outlined below in Section 11-27.

(8) Rough Shaft: With the machine at rest, remove the bearing cap and examine the shaft and bearings. If either the shaft or the bearing itself is rough, use a fine file to smooth the roughness.

26. Hot Field Coils. Heating of the field coils is usually due to one of the following causes:

274 Essentials of Electricity

Test for and correct   1. Too large field current
in this order: 2. Moisture in windings

(1) To Test for Excessive Field Current:
FIRST: Feel each of the coils. If they are all hot, the field current must be reduced, usually by changing the setting of the field rheostat. If any coils are cool, these coils are probably short-circuited.

SECOND: As a further check, measure the voltage across each coil. If any coil has a much lower voltage drop than that of the other coils, take off this coil and examine it for a short circuit.

(2) To Test for Moisture in Windings: Note whether or not the coils steam or feel damp to the touch. If the coils seem damp, set up about three-quarters of the normal current in the field circuit; allow the coils to dry with this heating for several hours.

27. Hot Commutator. A hot commutator may be due to one of the following causes:

Test for and repair   1. Near some hotter part of machine
in this order: 2. Sparking under brush
                               3. Poor brush contact

(1) Test for Hotter Part: Place your hand on the bearing near the commutator. If this is hotter than the commutator, heat comes from the bearing, and it should be checked as outlined in Section 11-25 above.

(2) Test for Sparks Beneath Brushes: Sight between the brush and commutator to see if sparks are occurring. If there are sparks, apply the tests of Section 11-23 above to locate and correct the problem.

(3) Test for Poor Brush Contact:
FIRST: Test the tension of the brushes by pulling up one after another (being careful not to open the circuit); correct any defects.

SECOND: With the machine stopped, take out each brush and examine its contact surface to see if the brush has its entire face bearing on the commutator. If not, sandpaper the high spot down and fit the surface to the commutator as directed in Section 11-23 above.

THIRD: If the commutator seems dry, try application of a commutator lubricating compound; these compounds are often available in easily applied sticks.

28. Hot Armature Coils. When there is an odor of hot insulation around a machine, it is always good practice to stop the machine and feel the armature coils. Heating may be due to some of the following causes:

Test for and correct as 1. Overload
indicated below: 2. Damp windings
　　　　　　　　　　 3. Short-circuited coils

(1) To Test for Overload: See instructions in Section 11-23 above.

(2) To Test for Dampness in Coils: Look for any steaming coils. With the machine at rest, feel the coils for dampness. If any are damp, arrange nearly full-load current in the armature circuit only; turn the armature slowly for several hours while the dampness bakes out.

(3) To Test for a Short-Circuited Coil: See the instructions for finding a short-circuited coil, given above in Section 11-23.

29. Generator Fails to Build Up. This trouble is always due to the failure of the magnetic flux lines of the field to "build up," and the cause may usually be found in one of the categories listed below:

Test for and correct 1. Field connections reversed
in the following 2. Wrong direction of rotation
order: 3. Speed too low
　　　　 4. Field circuit open
　　　　 5. Not enough residual magnetism
　　　　 6. Machine short-circuited

(1) Test for Reversed Field Connections: Reverse the connections of the field to the armature. (Review Section 11-12 for a discussion of the theory involved in this change.) If this reversal does not cause the voltage to rise to the normal value, replace the field connections as before.

(2) Test for Direction of Rotation: Reverse the direction of rotation of the armature. If this change does not correct the trouble, change back again to the former direction.

(3) Test for Too Low Speed: Measure the speed of the shaft with an appropriate speed counter or tachometer, and compare the measured value with the rated speed of the machine for operation as a generator. Try a slight increase in speed.

(4) Test for an Open-Field Circuit:
FIRST: Stop the machine and disconnect the armature. Establish a current in the field from a battery or other source. The inductive effect of the field should create a noticeable spark when the circuit is broken. If no continuity is indicated, or if no spark occurs when the circuit is opened, there is a break in the field circuit.

SECOND: Using similar continuity checks, determine where in the field circuit the break is located. (See also the weak-field tests of Section 11-

23 above.) If there is a break inside any coil, the coil must be repaired or replaced.

(5) Test for Low Residual Magnetism: Arrange the field winding so that it may be separately excited from a suitable source (dry cells or an automotive battery, if no other source is available). Observe whether the separate excitation increases or decreases the terminal voltage of the armature. If the voltage goes down with separate excitation (or reverses polarity), try changing the polarity of the external source. As noted in Section 11-12, some generators may occasionally fail to build up unless sufficient residual magnetism has been provided by an external source temporarily connected for this purpose.

(6) Test for Short Circuit of the Machine:

FIRST: Disconnect the load on the generator by opening the line switch or circuit breaker. If the generator now builds up its voltage, reclose the load switch. If the terminal voltage does not fall excessively, the failure to build up properly was that the load resistance was too small (load current too high) when the building-up process was started. If there is a short circuit in the line wires or in the load, the circuit breaker or fuses will disconnect the machine. The short circuit should be located as outlined in Section 11-23 above, before the circuit is reclosed again.

SECOND: If the machine fails to build up with the line switch open, there is probably a short circuit at or inside of the machine. Stop the machine and place a strip of paper under all the brushes, without otherwise altering the connections. With an appropriate source and ammeter, determine the current supplied by the source to the generator terminals, with the line switch off. If the ammeter reads more than the small current required by the field coils, there is an unwanted short circuit inside the machine. A common cause for such a short circuit is that some of the terminals or connections have become "grounded" (or connected) to the frame of the machine. The following steps outline a scheme for determining where these faulty connections are located.

THIRD: Make a careful inspection of all leads and terminals for evident grounding. If no places can be seen, disconnect the armature from the field. With a low-range voltmeter and some dry cells (or any convenient continuity checker), observe whether there is a path for current from the field to the frame. Remove the brushes and test between the frame and brush holders for a stray path. The same test can be made between a commutator segment and the frame. Before removing any coils for repair, make a second check to be sure that the fault does not lie at the terminals themselves.

11: D-C Machines

30. Voltage of Generator Too Low. Most of the causes listed below come under the heading of a weak field:

Test for and correct in this order:
1. Too much resistance in field
2. Overload
3. Speed too low
4. Some poles reversed or short-circuited

(1) Too much Resistance in the Field:
FIRST: Gradually cut out all resistance of the field rheostat.

SECOND: Look for a loose or corroded field connection.

(2) Test for Overload: See the first part of Section 11-23 above.

(3) Test for Too Low Speed: This test is given above in Section 11-29.

(4) Test for Reversed or Short-Circuited Field Coils: Use a compass to check polarity of poles, as outlined in Section 11-23 above. The weak-field test of Section 11-23, or the test outlined in Section 11-26 may be used to locate a shorted field coil.

31. Generator Voltage Too High. The usual causes are:

Test and correct in the following order:
1. Too strong field
2. Speed too fast

(1) Test for Too Strong Field: Gradually cut in all of the resistance of the field rheostat.

(2) Test for Speed: Measure the speed with a tachometer or other speed counter, and compare with the speed rating for the machine. If possible, lower the speed.

32. Motor Fails to Start. If a motor with a manual starting box does not start when the arm of the box has been advanced to its third point, open the line switch or circuit breaker and then release the arm of the box. Failure to start may be due to the following causes:

Test for and correct in the order given:
1. Wrong connections
2. Open circuits in connecting wires
3. Field weak
4. Overload
5. Friction excessive

(1) Wrong Connections: Go over all connections carefully, being sure that they agree with the diagrams and principles given in the earlier part of this chapter. Note especially the connections to the starting box or automatic starter. If possible, arrange to start the motor without a load.

(2) Test for Open Circuits:

FIRST: With the line switch closed, place a voltmeter across the line between the fuses and the machine. If there is no reading, look for blown fuses, open circuit breakers, etc.

SECOND: Put a voltmeter across the armature terminals, and try starting the motor as before; do not move the manual starting arm beyond the first three points if the motor does not start. If there is no reading on the voltmeter, there is either a wrong or a broken armature connection.

THIRD: Put a voltmeter across the field terminals, and again try to start the motor; do not move the manual starting arm beyond the first three points if the motor does not start. If the voltmeter gives no indication, there is either a wrong or a broken field connection. If, in these tests, the voltmeter reads the apparently correct values, proceed with the following tests.

(3) Test for Weak Field: Cut out all resistance in the field rheostat.

(4) Test for Overload: Remove the load and try starting the motor. Put on the load slowly. If the fuses or circuit breakers blow, the load is too heavy.

(5) Test for Friction:

FIRST: With an ammeter correctly connected to read the line current, start the motor with its load removed. If the current observed for this no-load running condition is more than 6 or 8% of the normal full-load current, there is too much friction present.

SECOND: Continue the tests outlined in Section 11-23 above for sparking brushes due to overload.

33. Motor Speed Too High. Overspeeding of a motor is usually due to a weak field:

Test for and correct in the order given:
1. Too much field-rheostat resistance
2. Connections wrong
3. Open field circuit

(1) Test for Rheostat Setting: Cut out all the resistance of the shunt-field rheostat.

(2) Test for Wrong Connections: Go over all connections to be sure that they agree with the diagrams and principles discussed in the earlier sections of this chapter.

(3) Test for Open Field Circuit: An open field circuit makes an unloaded shunt motor race and spark badly. With the armature leads disconnected, connect the field terminals to an appropriate source of voltage. If there is no spark when the field circuit is broken, there is an open

circuit in the field. For locating the break in the circuit, see *To Test for Weak Field* in Section 11-23 above.

34. Motor Speed Too Low. Look for the following causes of low motor speed:

Test for and correct in the order given below:
1. Overload
2. Too little field resistance
3. Excessive friction
4. Short circuit inside armature

(1) Test for Overload: Follow the outline of testing for overload, as given in Section 11-23 above.

(2) Test for Rheostat Setting: Cut in more resistance on the field rheostat.

(3) Test for Excessive Friction: An outline is given in Section 11-32 above.

(4) Test for Short Circuit in Armature: Review "To Test for Faulty Armature Coils" in Section 11-23 above.

35. Speed Control of Shunt Motors. Two important characteristics of the shunt motor are:

(1) Nearly constant speed at all loads.

(2) Speed can be controlled by added field resistance.

In Section 11-16, a greatly weakened field in a shunt motor was seen as the cause of an overspeed danger. However, a rheostat inserted in series with the small field current of the shunt motor can be used to increase the speed of the motor somewhat, without encountering dangerously high speed.

If a shunt motor is operated with no added resistance in its field circuit, the speed that its shaft develops at normal line voltage is known as *base speed*. When resistance is added to the field circuit with the field rheostat, the speed of the motor *increases* above the value of base speed. Whether the shunt motor has light load or full load on its shaft, the speed will remain reasonably constant, once the field-rheostat setting has been made.

Resistance may be added to the armature circuit of a shunt motor to cause operation at speeds *below* base speed. However, for a reduction in speed to one-half of base speed, about half of the nominal power of the motor must be wasted in this added resistance. With added resistance in the rotor circuit, the speed of the shunt motor no longer holds constant for a fixed setting of the field rheostat. Because of both of these drawbacks, control of the speed of a shunt motor by added armature resistance has limited application.

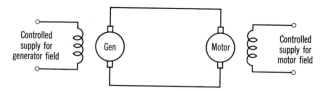

Fig. 232. Circuit of separately excited motor for speed control.

For traction or hoisting applications, the *series* motor is the usual choice of motor. Added resistance in series with the entire motor circuit is often a simple and convenient means for controlling the speed of a series motor.

36. Speed Control with Separate Excitation. Many rolling and winding operations in industry require careful speed control. If each motor has its own generator to supply its armature circuit, while each field circuit is supplied from a separate source, good control of speed below base speed is possible. Figure 232 illustrates the essential form of this separately excited arrangement, sometimes also called the *Ward Leonard connection.*

Of course, the need for this specialized speed control must be great to justify the cost: there must be a driving machine for the d-c generator, the separate d-c generator, and the separately excited d-c motor for performing the actual work at controlled speed, together with important associated starting devices and controllers. However, little power is wasted if the working d-c machine is operated with full field current below base speed. For speeds above base speed, the d-c generator is caused to apply normal full voltage to the armature circuit of the working motor, and the magnetic field is weakened on the motor, often with the added resistance of a field rheostat.

37. Control Generators. When the motor to be controlled in the separately excited circuit of Section 11-36 is very large, the control over its generator may not be easy to accomplish. In such a circumstance, the field current of the generator may be supplied from a *pilot generator,* as shown in Fig. 233. The operator of the equipment may then set the field rheostat for the pilot generator, which then in turn establishes the correct field current for the main generator, this main generator then supplying the correct armature voltage to the main motor.

In a single separately excited d-c generator, there is an *amplification* of effect, the operation of the armature circuit involving more power than the operation of the field circuit. For example, the armature of a

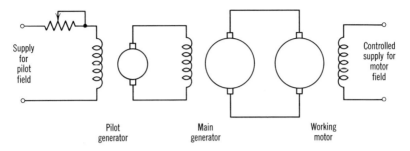

Fig. 233. The pilot generator affords delicate control over the separately excited motor.

d-c generator may supply its load with a 50-ampere current, while the field rheostat may control only a 2-ampere current. From field to armature, we may say that there is an amplification of current of **25**. If we had computed the appropriate values of power in watts, we could determine the *amplification of power* from the field to the armature.

It is important to note that the presence of amplification does not mean that the generator is working on a "something-for-nothing" principle. The useful power output comes from the driving machine through the energy-conversion process in the generator, as described in Section 11-15. The power in the armature is under the control of the power in the field. Comparison of the output power from the armature with only the input power to the field leads to the idea of amplification.

The amplification of the d-c generator makes possible delicate control over a separately excited motor. The motor in Fig. 232 can be controlled in speed (or in torque for a given load) with only the small effort needed to adjust the rheostats in the field circuits of the motor and generator. With one or more pilot generators arranged in cascade to produce further amplification, the control of large power may be executed with extremely delicate devices; one ultimate design goal could be to make the control process automatic.

For automatic control, time changes become important. There must be only a short time between the moment when a current is set at its correct value in the field of a pilot generator and the later moment when the speed of the working motor reaches the desired value. Special-purpose pilot generators for large power amplification and rapidly forced outputs are known as *control generators*. One design, having the equivalent amplification of two pilot generators and having much shorter time lag than two separate generators, is called an *Amplidyne*. Figure 234 shows a particular form of this type of control generator.

38. Conclusion. Consideration of the amplification of a d-c generator

282 Essentials of Electricity

Fig. 234. This standard Amplidyne is often employed as the pilot generator in the circuit of Fig. 233. *General Electric Company.*

focuses attention on the ability of the machine to transmit through itself a useful action, i.e., the turning of a field rheostat sets up a useful output power in the armature circuit. Another way of describing this process is to say that the turning of the field rheostat represents a *signal* that some sort of action should take place. The d-c generator then processes this signal by releasing power that passes the information of the signal along to the next device (another generator or a motor) connected in cascade following the generator.

In this book attention has been directed to voltages and currents that are constant, or at least constant for long stretches of time. For systems of automatic control employing d-c machines, the useful signals change with time. Before the student can pursue a further study of such systems, he must begin a study of the behavior of circuits having time-varying voltages and currents.

SUMMARY

GENERATOR—a machine delivering electric power when mechanical power is put into it.

MOTOR—a machine delivering mechanical power when electrical power is put into it.

D-C MACHINE—a term which includes both motor and generator. The same machine may be used either as a motor or as a generator.

TWO BASIC PRINCIPLES—every d-c machine has the same two principles of operation: MOTOR ACTION, and FARADAY'S LAW.

MOTOR ACTION—whenever a current-carrying wire is immersed in a magnetic field, there is a force that tries to move the wire.

11: D-C Machines

FARADAY'S LAW—when a wire that is oriented at right angles to a magnetic flux is moved parallel to itself and at right angles to the flux, a voltage is generated in the wire; the size of the voltage depends upon the length of the wire involved with the flux, the strength of the magnetic flux, and the speed at which the wire moves.

STATOR designates the stationary part of a machine, and ROTOR designates the movable portion.

FIELD WINDING designates the collection of electromagnets in a machine that create the principal magnetic flux; in a d-c machine, the stator winding is called the FIELD.

ARMATURE means the rotor winding and rotor structure of a d-c machine; the winding is embedded or "armored" in iron.

The size of the FARADAY VOLTAGE in the armature conductors changes with time.

The COMMUTATOR and BRUSHES form a switching arrangement that causes the time-changing Faraday voltages to appear as a constant direct voltage at the armature terminals.

FIELD CONNECTIONS may be arranged to be either: SEPARATELY EXCITED—field current supplied from an external source. SELF EXCITED—(for a generator) field current supplied from power produced in the machine itself; motor connections are usually classified this way.

SELF-EXCITED CONNECTIONS are classified: SHUNT—a stator winding (shunt field) is connected in parallel with the rotor-circuit terminals. SERIES—a special main-field winding carries the entire rotor-circuit current in a series connection. COMPOUND—a shunt winding supplies the principal flux, and a series winding adds a modifying effect.

The POLES should normally be arranged alternately north and south around the frame; their strength depends upon the number of AMPERE-TURNS in the coils.

COMMUTATING POLES (or INTERPOLES) are small poles, located between the main poles, for the purpose of preventing sparking at the brushes. Their coils are always in series with the rotor circuit.

The NUMBER OF BRUSHES (or brush sets) is usually the same as the number of main poles. All (+) brushes are joined in parallel, and all (−) brushes are joined in parallel.

An EQUIVALENT CIRCUIT for the rotor circuit of any d-c machine has an equivalent internal voltage in series with an equivalent internal resistance.

BUILDING UP describes the process, in a self-excited generator, by which the low voltage of residual magnetism is enhanced until the machine produces its normal voltage.

RULES FOR CONNECTING A MACHINE FOR SERVICE. Trace out the leads to the terminals, making a clear diagram showing all the connections that are needed. Follow the diagram carefully in making the connections.

TORQUE is a description of the turning effort of a rotary device; in a d-c machine, torque is produced as a result of the forces of motor action on the current-carrying conductors of the rotor circuit.

HORSEPOWER equals torque in pound-feet times speed in rpm, all divided by 5250.

ENERGY CONVERSION describes the process whereby electric energy is interchanged for mechanical energy (or vice versa) in the rotor of a d-c machine.

The BACK VOLTAGE of a motor is its internal generated voltage.

284   Essentials of Electricity

STARTING RESISTANCE must be added to the armature circuit of a motor, until the back voltage has become sufficiently large; the internal resistance of most motors is very low.

A STARTING BOX contains a resistance to be added to the armature circuit of a motor during starting, and a manually operated switch arm controlling the amount of resistance connected.

$$\text{CURRENT (in the armature)} = \frac{\text{LINE VOLTAGE} - \text{BACK VOLTAGE}}{\text{RESISTANCE (of armature circuit)}}$$

NO-FIELD RELEASE designates an electromagnet on a manual starting box that shuts a shunt motor down if the field current becomes too small for safe operation.

NO-VOLTAGE RELEASE designates an electromagnet on a manual starting box that shuts the motor down if the line voltage falls below a certain value.

A THREE-POINT STARTING BOX for shunt motors contains a starting resistance and a no-field release; there are three terminals on the box; Field, Armature, and Line; the remaining field and armature terminals of the motor must be connected to the remaining terminal of the line.

A FOUR-POINT STARTING BOX has four terminals and a no-voltage release. Since the release coil is connected across the line terminals, both line wires must be brought to this box; this extra line terminal is connected to the line on the same side as the common armature and field connection of the motor.

AUTOMATIC STARTERS employ magnetic contactors and relays to insert starting resistance in the armature circuit at the proper time.

SERIES MOTORS find useful application for hoisting and traction work. For two series motors, the method of series-parallel starting allows each motor to act as part of the starting resistance for the other.

TO REVERSE THE DIRECTION OF ROTATION of a shunt motor, reverse the direction of the current in EITHER the field or the armature, NOT IN BOTH.

CAUTION. Do not open the field circuit of a shunt motor.

Do not start an unloaded series motor or take off the load of a series motor while it is running.

Do not stop a motor by pulling back the arm of its starting box. Open the main switch or circuit breakers.

Do not allow the arm of a box designed "for starting duty only" to remain on an intermediate tap point.

Note carefully the SIGNS of trouble. Make tests for the CAUSES of trouble in the order given in Section 11-22. In this way one cause after another is eliminated until the final cause is removed.

TO AVOID TROUBLE:

First: Keep the commutator clean and smooth, and keep the brushes well fitted to the commutator.

Second: Keep all parts of an electric machine dry.

PROBLEMS

Draw diagrams of the electrical connections before making computations.

Prob. 33-11. A shunt generator having a terminal voltage of 120 volts supplies line wires with a current of 42 amperes. The field coils have a resistance of 180 ohms.

(a) What is the current in the field winding?
(b) What is the current in the armature?

Prob. 34-11. What is the electric output power of the generator of Prob. 33-11?

Prob. 35-11. How much power, in watts, is required to excite the shunt field of the generator in Prob. 33-11?

Prob. 36-11. The generator of Prob. 33-11 has an armature resistance of 1.2 ohms (equivalent internal resistance, including brushes). What power is lost in this resistance?

Prob. 37-11. For the generator of Prob. 36-11, what is its internal generated voltage?

Prob. 38-11. For the generator of Prob. 37-11, how much power (in kilowatts) is converted from mechanical to electric form in the internal energy-conversion process? Also express this amount of power in horsepower.

Prob. 39-11. The short-shunt compound generator of Fig. 213 maintains 120 volts across the terminals of the five lamps. The series field has a resistance of 0.25 ohm, and the shunt field has a resistance of 250 ohms. Each lamp requires a current of 2 amperes.

(a) What is the size of the current in the shunt field?
(b) Determine the current in the series field.
(c) Compute the current in the armature.

Prob. 40-11. How much power is lost in the series field of the generator of Prob. 39-11?

Prob. 41-11. How much power is lost in the shunt field of the generator of Prob. 39-11?

Prob. 42-11. The resistance of the armature of the generator of Prob. 39-11 is 0.24 ohm. What is the voltage drop across this internal resistance of the rotor circuit?

Prob. 43-11. How much power is lost in the armature resistance of the generator of Prob. 42-11?

Prob. 44-11. What is the Thevenin equivalent voltage in the armature (internal generated voltage) for the machine of Prob. 42-11?

Prob. 45-11. For the energy-conversion process in the armature of the machine in Prob. 42-11, compute the mechanical power in horsepower that is converted into electric power.

Prob. 46-11. If the generator of Prob. 39-11 were reconnected for long-shunt operation, find the answers to parts (a), (b), and (c) of Prob. 39-11 for the new connection.

Prob. 47-11. How much power is lost in the series field of the generator of Prob. 46-11?

Prob. 48-11. How much power is lost in the shunt field of the generator of Prob. 46-11?

Prob. 49-11. Is the shunt generator in Fig. 235 connected to the line properly? If not, show the proper connections.

Prob. 50-11. Is the shunt generator of Fig. 236 connected properly? If not, show the proper connections in a new diagram.

Prob. 51-11. When the generator of Fig. 236 is properly connected, assume that the upper line wire is the positive (+) wire. What changes would you make to cause the upper line wire to be the negative (−) wire?

286   Essentials of Electricity

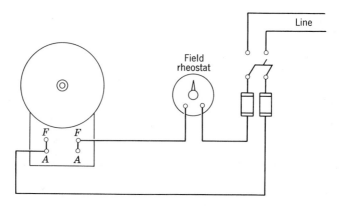

Fig. 235. Diagram of connections for a shunt generator.

Prob. 52-11. Is the shunt generator in Fig. 237 correctly connected? If not, show the necessary changes in a new diagram.

Prob. 53-11. In the compound generator of Fig. 238, $F1$ and $F2$ are the shunt-field terminals, and $S1$ and $S2$ are the series-field terminals. Are the connections right for long-shunt operation? If not, show all the necessary changes in a new diagram.

Prob. 54-11. Show the correct connections for the machine of Fig. 238 when it is arranged for operation as a short-shunt compound generator.

Prob. 55-11. Show the connections for the machine of Fig. 238 for operation as a simple shunt generator.

Prob. 56-11. If the shunt generator of Fig. 236 failed to build up after it was

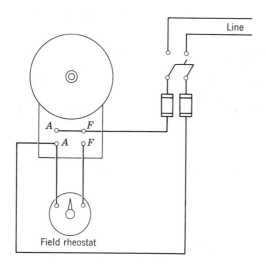

Fig. 236. Note the connection of the field rheostat.

connected according to a correct wiring diagram, what change in the connections would you make?

Prob. 57-11. Show the correct connections for the machine of Fig. 235 for operation as a motor.

Prob. 58-11. Make the necessary changes in your diagram of connections for Prob. 57-11 to reverse the direction of rotation of the motor.

Prob. 59-11. Show the connections for the generator of Fig. 236 for operation as a motor. Use a three-point box.

Prob. 60-11. Show the necessary changes in your diagram for Prob. 59-11 to reverse the direction of rotation.

Prob. 61-11. Sketch the connections for the generator of Fig. 237 for operation as a motor. Use a four-point box.

Prob. 62-11. Show the necessary changes in your connections for Prob. 61-11 to reverse the direction of rotation.

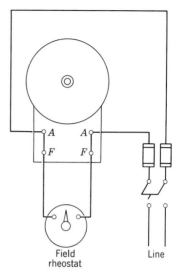

Fig. 237. Proposed connections for a shunt generator.

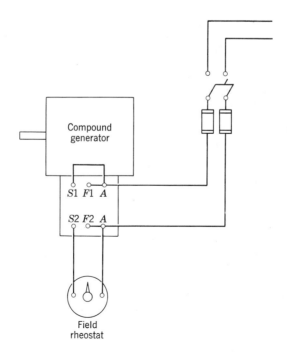

Fig. 238. Connections for a compound generator.

288 Essentials of Electricity

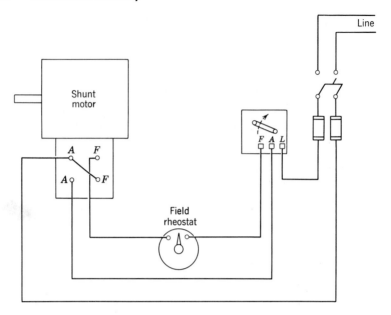

Fig. 239. Shunt motor with rheostat and starting box.

Prob. 63-11. Sketch the connections for the generator of Fig. 238 for operation as a shunt motor. Use a four-point box.

Prob. 64-11. If the mmf's of the shunt and series fields are additive when currents are directed in at *both* the $F1$ and the $S1$ terminals, show the correct connections for operating the generator of Fig. 238 as a long-shunt cumulatively compounded motor. Use a four-point box.

Prob. 65-11. Is the shunt motor of Fig. 239 connected correctly? If not, show the necessary changes on a new diagram.

Prob. 66-11. Make the necessary changes in your diagram of connections for the motor of Fig. 239 to reverse the direction of rotation.

Prob. 67-11. Make a diagram showing a 240-volt shunt motor connected through a four-point box and field rheostat to a three-wire system having 120 volts each side of neutral.

Prob. 68-11. In a new diagram, show the changes in connections for your diagram of Prob. 67-11 so that the shunt field will be excited from a 240-volt connection while the armature is supplied from a 120-volt connection.

Prob. 69-11. Show the correct connections for causing the reverse direction of rotation for Prob. 68-11.

Prob. 70-11. If the normal speed of operation of the motor in Prob. 67-11 is 1200 rpm, use Faraday's law to estimate the normal operating speed of the motor in Prob. 68-11.

Prob. 71-11. Make a sketch of the magnetic yoke shown in Fig. 240. On this sketch show a few of the magnetic flux lines for the main poles; be sure these paths are closed loops. Add a closed loop of flux for each commutating pole; label each interpole with its correct polarization (north or south).

Prob. 72-11. For the yoke of Fig. 240, show a winding on each polar projection; label each winding with the correct direction of current for motor operation in the direction shown.

Prob. 73-11. If the machine sketched in Fig. 240 is to run as a self-excited shunt generator, show: (*a*) polarization of the commutating poles, and the direction of rotation of the armature; (*b*) closed-loop paths of magnetic flux through each polar section of the yoke; (*c*) direction of current in the windings on the poles; and (*d*) connections of pole windings to the armature and the external circuit.

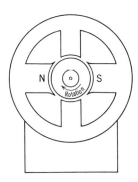

Fig. 240. Diagram of the pole structure of a 2-pole motor having commutating poles.

Prob. 74-11. What size rubber-covered copper wire must be run from the generator to the load in Prob. 33-11, in order to satisfy the requirements of the National Electrical Code?

Prob. 75-11. If the distance between the generator and the load in Prob. 74-11 is 400 feet, what is the total voltage drop in the line when the generator is delivering its full-load power?

Prob. 76-11. If it is desired that the drop along the 400-foot line of Prob. 75-11 shall be cut in half, what size copper wire should be used?

Prob. 77-11. What size rubber-covered aluminum wire should be used in Prob. 74-11?

Prob. 78-11. How much power is lost in the 400 feet of line in Prob. 75-11?

Prob. 79-11. If the same power were transmitted over the same line in Prob. 75-11 at *double* the sending-end voltage, how many watts would be lost in the line?

Prob. 80-11. The line voltage in Fig. 241 is 115 volts. The resistance of shunt field F is 200 ohms; the resistance of magnet coil M is 20 ohms. In

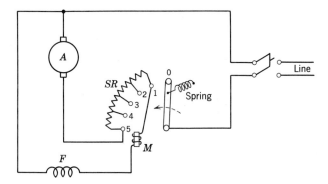

Fig. 241. If the current in field F should become zero, the spring returns the movable arm to its OFF position.

290 Essentials of Electricity

starting resistance SR, the resistances are: from point 1 to point 2, 5 ohms; from point 2 to point 3, 4 ohms; from point 3 to point 4, 3 ohms; and from point 4 to point 5, 1 ohm. The armature resistance is 2 ohms. After the main switch is thrown on, the starting arm is moved to point 1. Before the motor accelerates:

(*a*) What is the current in the armature?
(*b*) What is the current in the field?

Prob. 81-11. When the armature of the motor in Prob. 80-11 has attained enough speed to set up a back voltage of 25 volts, the starting arm is moved to point 2. Before the motor accelerates any further:

(*a*) What is the current in the armature?
(*b*) What is the current in the field?

Prob. 82-11. When the back voltage of the armature in Prob. 80-11 has reached 48 volts, the starting arm is moved to point 3. Before further acceleration:

(*a*) What is the current in the armature?
(*b*) What is the current in the field?

Prob. 83-11. After the armature of Prob. 80-11 has attained a back voltage of 102 volts, the starting arm is moved to point 4. For no further increase in speed, answer (*a*) and (*b*) of Prob. 80-11.

Prob. 84-11. When the armature of Prob. 80-11 attains a back voltage of 106 volts, the starting arm is moved to point 5. For no further increase in speed, answer (*a*) and (*b*) of Prob. 80-11.

Prob. 85-11. At a certain speed when the starting arm in Prob. 80-11 is on point 4, the line current of the motor becomes 3 amperes. What is its back voltage under these conditions?

Prob. 86-11. For the operating conditions of Prob. 85-11, what is the power in watts that is being converted from electric to mechanical form in the armature of the motor?

Prob. 87-11. A certain shunt motor has a line current of 80 amperes when operated under load from line wires at 115 volts. The resistance of the armature is 0.04 ohm, and the resistance of the field is 60 ohms.

(*a*) What is the current in each circuit?
(*b*) What power is lost in the field winding?
(*c*) What power is lost in the armature resistance?
(*d*) What is the total of parts (*c*) and (*d*)?

Prob. 88-11. For the motor in Prob. 87-11:

(*a*) What is the total power taken from the line?
(*b*) Why is this total power greater than the total lost power computed in part (*d*) of Prob. 87-11?

Prob. 89-11. A certain shunt generator has field windings with 2000 turns per pair of poles. What is the magnetomotive force per pair of poles due to the field winding of this generator when the field current is 0.70 ampere?

Prob. 90-11. A shunt motor has 800 turns per pair of poles in its field winding. When the field current is 3.2 amperes, what magnetomotive force per pair of poles exists in the field yoke due to the field winding?

Appendix

Essentials of Electricity

Table A*   Abridged Wire Table, Standard Annealed Copper

AWG NO.	DIAMETER MILS (1)	AREA, CIRCULAR MILS (1)	OHMS PER 1000 FT. (2)	AWG NO.	DIAMETER MILS (1)	AREA, CIRCULAR MILS (1)	OHMS PER 1000 FEET (2)
0000	460.0	211,600	0.04901	21	28.5	812	12.8
000	409.6	167,800	.06182	22	25.3	640	16.2
00	364.8	133,100	.07793	23	22.6	511	20.3
0	324.9	105,600	.09825	24	20.1	404	25.7
				25	17.9	320	32.4
1	289.3	83,690	.1239				
2	257.6	66,360	.1563	26	15.9	253	41.0
3	229.4	52,620	.1971	27	14.2	202	51.4
4	204.3	41,740	.2485	28	12.6	159	65.3
5	181.9	33,090	.3134	29	11.3	128	81.2
				30	10.0	100	104
6	162.0	26,240	.3952				
7	144.3	20,820	.4981	31	8.9	79.2	131
8	128.5	16,510	.6281	32	8.0	64.0	162
9	114.4	13,090	.7925	33	7.1	50.4	206
10	101.9	10,380	.9988	34	6.3	39.7	261
				35	5.6	31.4	331
11	90.7	8230	1.26				
12	80.8	6530	1.59	36	5.0	25.0	415
13	72.0	5180	2.00	37	4.5	20.2	512
14	64.1	4110	2.52	38	4.0	16.0	648
15	57.1	3260	3.18	39	3.5	12.2	847
				40	3.1	9.61	1080
16	50.8	2580	4.02				
17	45.3	2050	5.05				
18	40.3	1620	6.39				
19	35.9	1290	8.05				
20	32.0	1020	10.1				

* Compiled from National Bureau of Standards Circular 31, 4th ed., 1956.

Note (1): Diameters are rounded off to the nearest 0.1 mil. For No. 10 and larger sizes, areas are rounded off to four significant figures; for all smaller sizes, only three significant figures are kept.

Note (2): Resistance values are stated for 20° C (68° F), rounded off as for areas.

Table B Resistance of Commercially Available Materials

COMMERCIAL MATERIAL	OHMS PER CIRCULAR MIL-FOOT
Silver	9.8
Copper, annealed	10.4
Copper, hard-drawn	10.7
Gold	14.6
Aluminum	17.0
Lohm[1]	60.
Permanickel[2]	100.
Manganin	290.
Chromax[1]	600.
Nichrome[1]	675.
Karma[1]	800.

Notes: 1—Registered Trademark, Driver-Harris Company
2—Registered Trademark, The International Nickel Co., Inc.

Essentials of Electricity

Table C* **Safe Current-Carrying Capacities**

AWG NO.	PORTABLE CORDS NEC(1)	BUILDING WIRE			RUBBER-INSULATED CABLES	
		RUBBER OVER COPPER NEC(2)	VARNISHED CLOTH OVER COPPER NEC(3)	RUBBER OVER ALUMINUM NEC(4)	IPCEA(1)	IPCEA(2)
18	7					
16	10					
14	15	15	25		25	30
12	20	20	30	15	31	40
10	25	30	40	25	41	50
8		40	50	30	54	65
6		55	70	40	70	90
4		70	90	55	91	115
3		80	105	65	—	—
2		95	120	75	120	150
1		110	140	85	138	175
1/0		125	155	100	156	195
2/0		145	185	115	180	225
3/0		165	210	130	205	260
4/0		195	235	155	235	295
Area†						
250		215	270	170	255	320
300		240	300	190	285	360
350		260	325	210	310	390
400		280	360	225	335	420
500		320	405	260	380	480

* Data for this table were provided by the Simplex Wire & Cable Co.
† Area in 1000's of circular mils.
NEC(1): Portable Copper Rubber-Insulated Rubber-Sheathed Cords, up to three conductors, maximum conductor temperature 140° F.
NEC(2): Copper Building Wire, Code Rubber Insulation, not more than three conductors in raceway, temperature 140° F.
NEC(3): Copper Building Wire, Varnished-Cloth Insulation, not more than three conductors in raceway, temperature 167° F.
NEC(4): Aluminum Building Wire, Code Rubber Insulation, not more than three conductors in raceway, temperature 140° F.
IPCEA(1): Three Single-Conductor Rubber-Insulated Copper Cables in Underground Ducts, 0 to 5000 volts, 100 percent load factor, copper temperature 140° F.
IPCEA(2): Same as IPCEA(1), except copper temperature at 194° F.

Index

Accelerating contactor, 262
A. C. S. R., 192
Added armature resistance, 279
Alternator, 131
Aluminum wire, 192
American Wire Gage (AWG), 187
Ammeter, 60
Ampere, 1, 4, 7
Ampere-hour, 162
Ampere-hour meter, 164
Ampere-turn, 44
Amplidyne, 281
Amplification, 281
Annunciator, 214
Area, circle, 181
   effect on resistance, 177
Armature, annunciator, 215
   buzzer, 47
   machine, 228
      equivalent circuit, 237
      hot coils, 267
      resistance added, 279
Arrow, current, 14, 51
Atom, 12
Automatic control, 282
Automatic starter, 261
Automotive battery, 159, 196
AWG, 187

B. & S., 187
Back voltage, 249
Balance, dynamic, 272
Balanced three-wire system, 198
Base speed, 279
Battery, 11, 131
   automotive, 159, 196
   cycling, 167
   dry-charged, 165
   floating, 167
   ideal, 14
   industrial-truck, 167
   storage, 159
Bearings, hot, 267
Bell, single-stroke, 211
   vibrating-stroke, 212
Biasing, 108
Bobbin, 57
Brake, eddy-current, 127
   permanent-magnet, 125
Bridge connection, 79
British units, 113
Brown and Sharpe Gage (B. & S.), 187
Brushes, 228
   set of, 233
   setting of, 271
   sparking at, 267
Build-up, failure to, 267

296 Index

Build-up, shunt-generator, 239
Bulb, three-way, 218
Buzzer, 47, 213

Calibration, 59
Capacity, current-carrying, 193, 294
 power-handling, 122
Causes of trouble, 266
Cell, 132
 dry, 132, 154
 nickel-cadmium, 170
 nickel-iron-alkaline, 169
 nuclear, 172
 secondary, 159
 silver-zinc, 171
 solar, 172
Cells, in series or parallel, 138
 series-parallel grouping, 151
Charge, 1
 flow, 2, 13
 negative, 12
Charging, 144, 164, 249
 constant-current, 166
 constant-potential, 166
 high-rate, 166
Circuit, 12
 electric, 3, 4, 11
 equivalent, 134, 145, 236
 ladder, 89, 103, 150
 short, 276
Circuit diagrams, 11, 83
Circuit symbols, 11
Circular mil, 182
Circular mil-foot, 182
Circumference, circle, 181
Commutating pole, 233
Commutation, 229
Commutator, 228
 action, 229
 hot, 267
Compass, 45, 50
Compass needle, 42
Compound connection (machine), 233
Compound-generator connection, 241
Compounding, cumulative, 233
 differential, 233
 flat, 242
 long-shunt, 242
 short-shunt, 242
Compound motor, reversal, 260
Concentric stranding, 191

Conductance, 30
Conductor, 176
Connection, compound, 233
 compound-generator, 241
 grounded, 276
 parallel, 17
 series (circuit), 17
 series (machine), 233
 series-parallel, 79
 shunt (circuit), 17
 shunt (machine), 232
 shunt-generator, 237
 Ward Leonard, 280
Constant of proportionality, 7
Constant-current charging, 166
Constant-potential charging, 166
Contact, holding, 211
 relay, 206
Contactor, 208
 accelerating, 262
 latching, 210
Control, automatic, 282
 incandescent lamps, 217
 series-parallel, 265
Control generator, 281
Controller, magnetic, 211
Conventional current, 12, 14
Conversion, energy, 245
Copper wire, 183
Copper wire table, 187, 292
Core, iron, 47, 57
Coulomb, 1, 12, 162
 joule per, 3, 111
 per second, 2
Counterelectromotive force, 249
Cranking ability, 163
Cross, 51
Cumulative compounding, 233
Current, 1, 13
 arrow, 14, 51
 conventional, 12, 14
 direction, 13
 distribution, 92
 eddy, 127
 electric, 13
 electron, 12
 measurement, 62
 normal discharge, 163
 Ohm's law, 10
 parallel circuit, 27
 series circuit, 18

Index

Current, voltmeter, 71
Current-carrying capacity, 193, 294
Current-carrying wire, 49
Current circuit, 119
Current law, Kirchhoff's, 80, 92
Current per volt, 30
Cycling battery, 167

D'Arsonval movement, 56
D-c generator, 131, 223
D-c machine, 223
D-c motor, 223
Dead-front construction, 256
Decrease in potential, 81
Deep discharge, 166
Deflection, full-scale, 62, 68
Destructive electrolysis, 158
Diagnosis, 266
Diagram, circuit, 11, 83
    voltage, 84
Diameter, circle, 181
Difference in potential, 4, 12
Differential compounding, 233
Direction, current, 13
Direction of rotation, 234
Discharge, deep, 166
    normal current, 163
Disconnect, 254
Distance (for work), 3, 111
Distribution, current, 92
Dot, 51
Double-pole switch, 11, 206
Drop, annunciator, 214
    line, 87, 120, 185
Drop in voltage, 81, 86
Dry cell, 132, 154
Dry-charged battery, 165
Dynamic balance, 272

Eddy current, 127
Eddy-current brake, 127
Efficiency, 122
Electric bell, 211
Electric circuit, 3, 4, 11
Electricity, 1
Electrochemical equivalent, 156
Electrode, 132
Electrolysis, 156
    destructive, 158
Electrolyte, 132
Electromagnet, 38, 44, 229

Electron, 12
Electron current, 12, 13
Electron tube, 12
Electroplating, 157
Electrotyping, 158
Elements, circuit, 12
Energy, 124
Energy conversion, 245
Equivalent resistance, 99
Equivalent linear representation, 137, 141
Equivalent resistance, parallel circuit, 31
Equivalent circuit, 134, 145, 236
    armature, 237
    stator, 236
    Thevenin, 146
    rotor, 237

Failure, to build up, 267
    to start, 268
Faraday's law, 223, 239
Faraday voltage, 237, 247
Field, 228
    magnetic, 43
    series, 231
    shunt, 230
Field coils, hot, 267
Field of force, 43
Field rheostat, 237
Field strength, magnetic, 53
Figure-of-merit, 67, 72
    reciprocal of, 68
Fixed-coil current, 119
Fixed-current load, 92
Fixed load resistance, 96
Fixed source voltage, 99
Flashover, 250
Flat compounding, 242
Floating battery, 167
Flow of charge, 2, 13
Flux, magnetic, 40
Flux lines, 38, 50
Foot-pound, 2, 111, 243
Force, 3, 111
    field of, 43
    magnetic, 38
    magnetomotive, 38, 44
Force of magnetism, 47
Four-point box, 259
Four-way switch, 219
Friction-and-windage losses, 246
Fuel (for cell), 154

Full-scale deflection, 62, 68
Fuse, 254

Generator, 15
  compound connection, 241
  control, 281
  d-c, 131, 223
  internal horsepower, 246
  pilot, 280
  separately excited, 237
  series, 243
  shunt build-up, 239
  shunt connection, 237
Generator regulator, 165
Grounded connection, 276

Half-scale resistance, 74
High-rate charging, 166
High speed of motor, 268
High voltage of generator, 267
Holding contact, 211, 261
  latching, 212
Horsepower, 112, 244
  generator internal, 246
Horseshoe magnet, 39, 56, 229
Hydrometer, 160

Ideal battery, 14
Ideal wires, 12, 14
Incandescent lamps, 11
  control of, 217
Increase in potential, 81
Industrial-truck battery, 167
Input, 122
Instrument, 55
  recording, 69
  watch-case, 76
Instrument springs, 57
Insulated Power Cable Engineers Association (IPCEA), 194
Insulator, 176
Internal generated voltage, 249
Internal power, 246
Internal resistance, 134
Interpole, 229
  polarity of, 234
IPCEA, 194
Iron core, 47, 57
Iron yoke, 206

Joule, 3, 111, 243
  per coulomb, 3, 111

Kilohm, 121
Kilometer, 121
Kilowatt, 121
Kilowatthour, 124
Kirchhoff's current law, 80, 92
Kirchhoff's voltage law, 81, 92

Ladder circuit, 89, 103, 150
  voltage, 96
Lamps, incandescent, 11
  control of, 217
Latching contactor, 210
Latching holding contact, 212
Latching relay, 210
Length, effect on resistance, 177
Level, voltage, 194
Lifting magnet, 48
Line drop, 87, 120, 185
  fixed-current load, 92
Line loss, 120, 194
Line wire, 86
Linear equivalent representation, 137, 141
Lines of flux, 38, 50
Load, fixed-current, 92
  fixed-resistance, 96
Local action, 154
Lock opener, electric, 214
Long-shunt compounding, 242
Losses, friction-and-windage, 246
Low speed of motor, 268
Low voltage of generator, 267

Machine, d-c, 223
  parts, 227
  overspeeding, 251
Machine troubles, 266
Magnet, horseshoe, 39, 56, 229
  permanent, 38
Magnetic controller, 211
Magnetic field, 43
Magnetic field strength, 53
Magnetic flux, rules, 40
Magnetic force, 38
Magnetic pole, 229
Magnetism, force of, 47
Magnetomotive force, 38, 44, 233
Main pole, 229
  polarity, 234
Main switch, 254
Manual reset, 216
Manual starting box, 254

Measurement, current, 62
   power, 115
   resistance, 71
   voltage, 65
Megohm, 9
Meter, 55
   ampere-hour, 164
   recording, 69
   watthour, 124
Metric units, 113
Mho, 30
Mil, 182
Mil-foot, circular, 182
Milliammeter, 56, 60, 63
Milliampere, 56
Milliampere-hour, 163
Millimho, 74
Millivolt, 56
Millivoltmeter, 64
Mmf, 233
Momentary push button, 210
Motor, 23
   d-c, 223
      failure to start, 268
      overspeeding, 254
      reversal, 259
      series, 264
      shunt, 248
      speed control, 279
      traction, 265
Motor action, 53, 57, 125, 131, 223, 247, 249
Movement, D'Arsonval, 56
   permanent-magnet moving-coil, 56
Moving-coil circuit, 119
Moving-coil movement, 56
Multimeter, 74
Multiplier, 64

National Bureau of Standards, 187
National Electrical Code (NEC), 194
NC (normally closed), 209
NEC, 194
Needle, compass, 42
Negative, 12
Negative terminal, 63, 132
Network reduction, 99
Neutral, 197
   open, 201
Newton, 111
Newton-meter, 243

Nickel-cadmium cell, 170
Nickel-iron-alkaline cell, 169
NO (normally open), 209
No-field release, 254
Noise, 267
No resistance, 12, 14
Normal condition of relay, 208, 261
Normal discharge current, 163
Normally closed, 208
Normally open, 208
North pole, of magnet, 39
No-voltage release, 259
Nuclear cell, 172
Nucleus, 12

Ohm, 7
Ohmmeter, 72
Ohm's law, 6, 24, 92
   current, 10
   resistance, 8
   voltage, 9
Ohms per volt, 67, 72
Open, normally, 208
Open neutral, 201
Operating coil, 215
Output, 122
Overspeeding machine, 251
Overspeeding motor, 254

Pair, polarity signs, 82
Parallel circuit, current, 27
   equivalent resistance, 31
   resistance, 28
   voltage, 26
Parallel connection, 17
   cells, 138
Parts, d-c machine, 227
Permanent magnet, 38
Permanent-magnet brake, 125
Permanent-magnet moving-coil movement, 56
Permeability, 39, 41, 47
Pilot generator, 280
Plus terminal, 66
Pointer, instrument, 57
Polarity, 13
   interpole, 234
   main pole, 234
   voltage, 82
Polarity rule, 13
Polarization, 164

300 Index

Polarization rule, 39
Pole, commutating, 233
  magnet, 39
  magnetic, 229
  main, 229
  salient, 230
Positive, 12
Positive terminal, 63, 132
Potential, 2
  series circuit, 18
  terminal, 16, 23, 33
Potential circuit, 119
Potential difference, 4, 12
Potential work, 3
Pound-feet, 243
Power, 112
  internal, 246
  measurement, 115
Power amplification, 281
Power balance, 116
Power-handling capacity, 122
Proportionality, 6, 24
  constant of, 7
Push button, 209
  momentary, 210

Rate, time, 111
Rating, 122
Ratio, 7, 111
Reciprocal, 30
  of figure-of-merit, 68
  resistance, 30
Recorder, 69
Recording instrument, 69
Recording meter, 69
Rectification, 131
Reduction, network, 99
Relay, 206
  latching, 210
  normal condition, 208
Relay contact, 206
Release, no-field, 254
  no-voltage, 259
Remedy, 266
Reset, electric, 216
  manual, 216
Reset button, 264
Reset coil, 215
Resistance, 7, 12
  added to armature, 279
  effect of area, 177

effect of length, 177
equivalent, 99
equivalent parallel, 31
half-scale, 74
internal, 134
measurement, 71
no, 12, 14
Ohm's law, 8
parallel circuit, 28
reciprocal of, 30
series circuit, 20
starting, 251
zero, 14
Resistance bridge, 79
Resistance of shunt, 62
Resistivity, 183
Resistor, 12
Reversal, motor, 259
Rheostat, field, 237
Right-hand rule, 44, 49
Rise in voltage, 81
Rope stranding, 191
Rotary switch, 218
Rotation, direction of, 234
Rotor, 225
Rotor circuit, Thevenin equivalent, 237
Rule, polarity, 13
  polarization, 39
  right-hand, 44, 49
Rules, magnetic flux, 40
  Thevenin method, 148
Runaway, series motor, 264
  shunt motor, 251

Salient pole, 230
Secondary cell, 159
Self-excited machine, 232
Semiconductor, 13, 132
Sense of voltage, 14
Sensitivity, 67
Separate excitation, speed control, 280
Separately excited generator, 237
Separately excited machine, 232
Series circuit, current, 18
  potential, 18
  resistance, 20
Series connection, cells, 138
  circuit, 17
  machine, 233
Series field, 231
Series generator, 243

Series motor, 264
  runaway, 264
Series ohmmeter, 72
Series-parallel cell grouping, 151
Series-parallel connection, 79
Series-parallel control, 265
Series resistance of multiplier, 65
Series resistor, 64
Set of brushes, 233
Setting of brushes, 271
Short circuit, 276
Short-shunt compounding, 242
Shunt, ammeter, 60
  resistance of, 62
Shunt connection (circuit), 17
Shunt connection (machine), 232
Shunt field, 230
  weakened, 251
Shunt-generator, build-up, 239
  connection, 237
Shunt motor, 248
  speed control, 279
Signs, pair of, 82
  plus and minus, 14
Signs of trouble, 266
Silver-zinc cell, 171
Single-pole switch, 11, 206
Single-stroke bell, 211
Solar cell, 172
Source, 14, 131
Source of work, 2, 14
Source voltage, 94
  fixed, 99
South pole (magnet), 39
Sparking at brushes, 267
Specific gravity, 160
Speed, base, 279
Speed control, separate excitation, 280
  shunt motor, 279
Speed too high, 268
Speed too low, 268
Springs, instrument, 57
Starter, automatic, 261
Starting box, 251
  manual, 254
Starting resistance, 251
Stator, 225
  equivalent circuit, 236
Storage battery, 159
Stranded wire, 191
Stranding, concentric, 191

rope, 191
Strength, magnetic field, 53
Switch, double-pole, 11, 206
  four-way, 219
  main, 254
  rotary, 218
  single-pole, 11, 206
  three-way, 218
  toggle, 206
Symbols, 11

Target, annunciator, 214
Terminal, negative, 63, 132
  plus, 66
  positive, 63, 132
Terminal potential, 16, 23, 33
Terminal voltage, 135
Thevenin equivalent circuit, 146
  rotor, 237
Thevenin method, rules, 148
Thevenin's theorem, 146
Three-point box, 255
Three-way bulb, 218
Three-way switch, 218
Three-wire system, 197
  unbalanced, 198
Timer, 262
Time rate, 111
Toggle switch, 206
Torque, 243
Traction motor, 265
Transistor, 13, 108
Trouble, causes of, 266
  signs of, 266
Troubles, machine, 266
Tube, electron, 12

Unbalanced three-wire system, 198
Unit-ampere method, 103
Units, British, 113
  metric, 113
Upscale, 58

Vibrating-stroke bell, 212
Volt, 3, 7, 111
Voltage, 3
  back, 249
  Faraday, 237, 247
  internal generated, 249
  ladder-circuit, 96
  measurement, 65

Voltage, Ohm's law, 9
  parallel circuit, 26
  polarity of, 82
  sense of, 14
  terminal, 135
  too low, 267
  too high, 267
Voltage diagram, 84
  unbalanced system, 199
Voltage drop, 81
Voltage law, Kirchhoff's, 81, 92
Voltage level, 194
Voltage rise, 81
Voltage source, 94
Voltmeter, 63
  current in, 71
Voltmeter-ammeter method, 71, 115
VOM, 73

Ward Leonard connection, 280
Watch-case instrument, 76

Watt, 112
Watthour, 124
Watthour meter, 124
Wattmeter, 118
Weak shunt field, 251
Winding, field, 228
Wire, 11
  aluminum, 192
  copper, 183
  current-carrying, 49
  ideal, 12, 14
  line, 86
  stranded, 191
Wire table, copper, 187, 292
Work, 2, 111, 243
  potential, 3
  source of, 2, 14

Yoke, iron, 206

Zero resistance, 14
Zero-set, 74